D1251844

Botanical Classifications

FRONTISPIECE

Portraits of Bentham and Hutchinson courtesy of the Hunt Institute for Botanical Documentation, Carnegie-Mellon University, Pittsburgh.

Portraits of Endlicher, Engler, and Melchior courtesy of the Hunt Institute, Pittsburgh, and the Berquis Botanical Garden, Stockholm.

Portrait of Hooker courtesy of the Hunt Institute, Pittsburgh, and the Linnean Society, London.

Portrait of Eichler courtesy of the Hunt Institute, Pittsburgh, and the Smithsonian Institution, Washington.

Portrait of Bessey courtesy of the Nebraska State Historical Society, Lincoln.

Portrait of Cronquist courtesy of the Keystone Press Agency and the New York Botanical Garden.

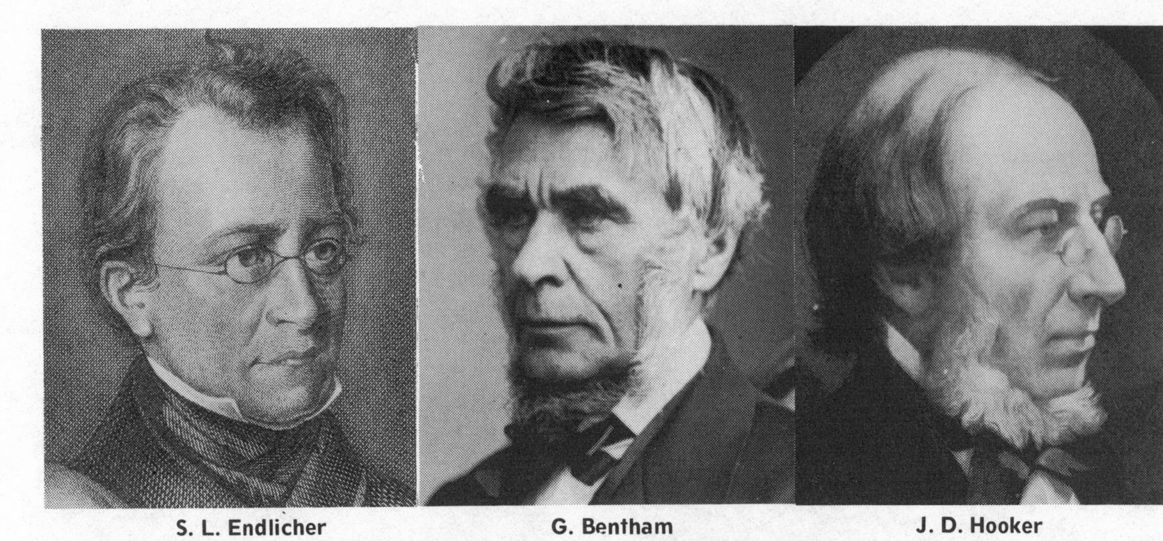

S. L. Endlicher — G. Bentham — J. D. Hooker

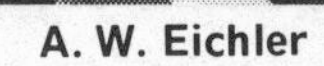

A. W. Eichler

A. Engler

C. E. Bessey

J. Hutchinson

H. Melchior

A. Cronquist

Botanical Classifications

a comparison of eight systems of angiosperm classification

BY

Lloyd H. Swift

Archon Books
1974

Library of Congress Cataloging in Publication Data

Swift, Lloyd H 1920-
Botanical classifications.

1. Botany--Classification. 2. Angiosperms.
I. Title.
QK95.S94 582'.13'012 74-12078
ISBN 0-208-01455-1

First published 1974 as an Archon Book
an imprint of
The Shoe String Press, Inc.
Hamden, Connecticut 06514

Printed in the United States of America

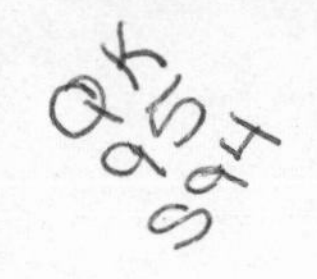

CONTENTS

CONTENTS

CONTENTS

PREFACE

All botanists and other persons studying plants must deal at least superficially with several systems of plant classification. Thus in Dewey classification libraries books are shelved according to the Bentham and Hooker system, with family names of this library classification following the Hutchinson system. Most floristic manuals and most herbaria are arranged by the Engler and Prantl system. Textbooks may follow the Bessey system while for phylogenetic considerations course lectures may follow the Cronquist system. A reference in all taxonomic courses is the _Syllabus der Pflanzenfamilien_, which follows the Melchior system. Thus there is presently a confusing mixture of systems in use.

Botanical Classifications gives orientation in this situation by comparing eight major systems of angiosperm classification on an order by order basis. It provides a focus for a central topic in botany--the classification of angiosperms. Previous writers have discussed groups of systems, but without attempting comparative analysis on an order by order basis. Kenneth M. Becker has published in _Taxon_ (1973) "A Comparison of Angiosperm Classification Systems" which features extensive tabular data, but lacks accompanying essay-type discussion.

PREFACE

One difficulty in carrying out comparative discussion is the large number of angiosperms and their many families and orders. This difficulty is here met by restricting the survey to the orders represented in the Central and Northeastern United States and to a selection of their families. A second difficulty is that the large number of proposed systems makes comparative discussion too lengthy for inclusion within a work of moderate size. This difficulty is met by restricting the discussion to eight representative systems. A third difficulty is semantic. Each of the eight systems differs from the others in terminology and nomenclature. This difficulty is met by translating the terminology and nomenclature of each of the eight systems into a single neutral terminology and nomenclature. The original terminology and nomenclature of the eight systems are, however, not lost sight of. They are presented in the form of glossaries accompanying the discussion for each order and higher category. The various names employed in the eight systems are indexed.

Additional features include general introductory materials, introductions to individual orders with etymologies for family names, plant formulae for representative families, illustrations of representative species within these families, and a general index which covers such terminology as requires special explanation.

Books presenting the eight systems are cited in the

introduction. Other works of special value in the preparation of the text include: N. L. Britton and A. Brown's _Illustrated Flora of the Northern United States_ ..., 2d ed., 3v., 1913, from which are derived the illustrations here used to accompany plant formulae; H. A. Gleason's _New Britton and Brown Illustrated Flora_, 3v, 1952, which has modern nomenclature and new descriptions of families; R. J. Pool's _Flowers and Flowering Plants_, 2d ed., 1941, which contains a system of plant formulae for families; P. H. Davis and J. Cullen's _Identification of Flowering Plants_, 1965, which gives for plant families brief descriptions that approximate plant formulae in their conciseness; C. L. Porter's _Taxonomy of Flowering Plants_, 2d ed., 1967, which includes floral diagrams comparable to floral formulae; and G. H. M. Lawrence's _Taxonomy of Vascular Plants_, 1951, which includes a listing of literature for each plant family of our area. Other surveys consulted include A. M. Johnson's _Taxonomy of the Flowering Plants_, 1931, and A. B. Rendle's _Classification of Flowering Plants_, 2d ed., v.1, 1956, 1st ed., v.2, 1925.

Although responsibility for the ultimate form of the text necessarily rests with the author, he gratefully acknowledges assistance from University of Nebraska botanists who read portions of the manuscript. Professor Wendell L. Gauger, Chairman of the Department of Botany prior to its assimilation into the School of Life Sciences, took an active interest in the project. Professors Robert C.

Lommasson and Robert B. Kaul gave attention to morphological aspects of the work. Professors John F. Davidson and Gregory J. Anderson gave attention to taxonomical aspects. The latter is currently on the staff of the University of Connecticut, Storrs. The author is grateful to his mother, Mrs. Grace H. Swift, for checking much of the manuscript on matters of form and style.

STATEMENT OF COVERAGE

Eight systems of angiosperm classification as they apply to the flora of the Central and Northeastern United States are here compared and contrasted. These are the system of Stephan Endlicher as given in his *Genera Plantarum* (1836 seq.), the system of George Bentham and J. D. Hooker as given in their *Genera Plantarum* (1862 seq.), the system of A. W. Eichler as given in *Blüthendiagramme* (1874 seq.), the system of Adolf Engler and K. A. E. Prantl as abstracted from their *Natürliche Pflanzenfamilien* by K. W. Dalla Torre and Hermann Harms in their *Genera Siphonogamarum* (1900 seq.), the system of C. E. Bessey as presented in a paper published in the *Annals* of the Missouri Botanical Garden (1915), the system of John Hutchinson as given in the second edition of his *Families of Flowering Plants* (1959) and in his *Evolution and Phylogeny of Flowering Plants; Dicotyledons: Facts and Theory* (1969), the system of Hans Melchior and his collaborators as given in the twelfth edition of the Engler *Syllabus der Pflanzenfamilien*, volume 2 (1964), and the system of Arthur Cronquist as given in his *Evolution and Classification of Flowering Plants* (1968). Most of these authors have employed systems differing somewhat from the ones selected, but such systems are largely excluded from consideration in the present survey. Many other systems, both similar and dissimilar to the specified systems, are also largely excluded. The systems included are fairly representative of taxonomic thought at the time they appeared, and all of them are or have been prominent and influential. The sequence of orders and families in this survey follows that of M. L. Fernald in the eighth edition of Gray's Manual of Botany. This is, with a few deviations, the sequence of the Engler and Prantl system as given by Dalla Torre and Harms.

The terminology for classification ranks, such as order, cohort, subphylum, etc. are quite variable from system to system. These differences constitute an important communication barrier, especially with respect to the understanding of the older systems. The nomenclature for several of the taxa are also variable from system to system. These differences are numerous for the older systems, and they are not rare for the more modern ones. In order to make the systems susceptible to comparative analysis it has been necessary to translate them into a consistent terminology for classification ranks and into a consistent nomenclature for the specific taxa. The third subsection within the present introduction discusses terminology, nomenclature, and synonymy. Glossaries are included within the basic sections for individual orders, and these specify the exact language of authors.

Introductory material provides a comparative discussion of higher categories in the eight systems. The 44 basic sections individually cover the classification of specific orders in the eight systems. These sections consist of four subsections. The first is introductory to the order. The second provides concise descriptions of selected families within the order by means of plant formulae. The third is a comparative discussion of the classification of the order in each of the eight systems. The fourth is a glossary for the original nomenclature of the authors of the eight systems.

O 2:1

THE EIGHT SYSTEMS

The system of Stephan Ladislaus Endlicher (1804-1849) appears in his *Genera Plantarum Secundum Ordines Naturales Disposita*, Vienna 1836-41, with supplements 1842-50. Stafleu's *Taxonomic Literature* pp. 130-31 provides collation for its 18 parts and 5 supplements. The signification of Endlicher's term *naturalis* is somewhat obscure, but in this connection it may be noted that Endlicher and Ungar in their *Grundzüge der Botanik* (1843) taught a doctrine of plant evolution. Endlicher's natural system progresses from putative simplicity to complexity. The system places the monocotyledons ahead of the dicotyledons, and it also places apetalous orders first in the classification of the dicotyledons. It is indebted to earlier systems, notably to that of the *Genera Plantarum* (1789) of Antoine Laurent de Jussieu, and it appears influential on the systems of Bentham and Hooker and of Eichler. Endlicher's work is the latest comprehensive survey of plant genera to be brought near completion by a single author.

The system of phanerogam classification by George Bentham (1800-1884) and Joseph Dalton Hooker (1817-1911) appears in their *Genera Plantarum ad Exemplaria Imprimis in Herbariis Kewensibus Servita Definita* (1862-83). Stafleu's *Taxonomic Literature* p. 28 provides collation for the seven parts of its three volumes. Lawrence's *Taxonomy of Vascular Plants* p. 118 provides a table specifying the precise division of authorship. The Bentham and Hooker system is based on the system of Augustin Pyramus de Candolle (1778-1841). De Candolle held that the basic flower types had radial symmetry, and that other forms were derived by abortion of parts, by degeneration of parts, and by adherence of parts. Thus in contrast with the de Jussieu system and its successors, such as the systems of Endlicher, Eichler, Engler, and Melchior, the de Candolle system and its successors, such as the systems of Bentham and Hooker, of Bessey, of Hutchinson, and of

Cronquist, lay stress on reduction in numbers of floral parts. These systems place or diagram the dicotyledons ahead of the monocotyledons, and they also place ranalian orders first in their classification of the dicotyledons.

The system of phanerogam classification used in the *Blüthendiagramme; Construirt und Erläutert* (1874-78) by August Wilhelm Eichler (1839-1887) accompanies the results of a fifteen-year analysis of floral morphology. In 1883 he published a revised version of his system in the third edition of his *Syllabus der Vorlesungen*. The *Blüthendiagramme* monocot system is closer to the Engler and Prantl system than is Eichler's 1883 system. The reverse, however, is true for the dicot gamopetalous orders. The two Eichlerian systems are almost identical for dicot non-gamopetalous orders. The Eichler *Syllabus* is essentially a mere list, whereas the *Blüthendiagramme* volumes contain, in addition to numerous floral diagrams, discussions on the various taxa.

The system of classification of the *Natürliche Pflanzenfamilien* (1887 seq.) edited by Heinrich Gustav Adolf Engler (1844-1930) and Karl Anton Eugen Prantl (1849-1893) is subordinate to its detailed descriptions of the individual families. Stafleu's *Taxonomic Literature* pp. 146-49 provides a collation of the first edition. It includes 33 pagination series with 177 publication dates. This complexity suggests the difficulties in maintaining rigid sequences of taxa within the many subdivisions of the multi-author work. In fact the so-called Engler and Prantl sequence of families is not the sequence that occurs in the *Natürliche Pflanzenfamilien*, but is that provided by the *Genera Siphonogamarum ad Systema Englerianum Conscripta* of Karl Wilhelm Dalla Torre (Latinized as C. G. Dalla Torre) and Hermann Harms which was published in parts between 1900 and 1907. The sequence of families as given in the Dalla Torre and Harms work has become standard for the arrangement of herbaria, and it is adopted, usually with more or less modification, in many floras. Robinson and Fernald in the seventh edition of Gray's *Manual of Botany* (1908) with scant modification adopt this system in place of the Bentham and Hooker system used by Gray in the sixth edition. Fernald in the eighth edition (1950) continues its use, but he raises several families to ordinal status, and in some cases this is sanctioned by Engler's later usage. In some cases Fernald rearranges the families within an order.

Although Prantl died in 1893, 31 years before the second edition of the *Natürliche Pflanzenfamilien* was commenced (1924), his name was retained on the work. Engler died in 1930, after several volumes of this edition had appeared. Collaborators continued the production until World War II not only brought this activity to a halt but destroyed (1943) the Berlin herbarium on which the work was largely based. This edition, so far as it has appeared, is organized with greater attention to the details of the classification of angiosperm taxa than is found in the first edition, but the system itself is little changed. Engler's later views are contained in the Engler and Gilg system as given in the single-volume ninth and tenth editions of the _Syllabus der Pflanzenfamilien_ (1924) and in the Engler and Diels system as given in the eleventh edition (1936) six years after Engler's death. The Engler and Prantl system is evolutionary in that it places taxa with putative primitive characters ahead of taxa with putative advanced characters. The system in general implies angiosperm evolution to have been from the morphologically simple to the morphologically complex, but many exceptions are recognized. Various synoptic keys to the Englerian system have been prepared.

The last of the several versions of the system of angiosperm classification of Charles Edwin Bessey (1845-1915) is outlined in the 1915 volume of the _Annals_ of the Missouri Botanical Garden under the title: "The Phylogenetic Taxonomy of Flowering Plants." Unlike the works of Endlicher, Bentham and Hooker, Eichler, and Engler in which the classification systems are more or less incidental to their main purposes, Bessey's paper is a simple listing of taxa arranged to specify putative phylogenetic sequences. That the system became widely influential is to be attributed not so much to the merit of the sequences themselves as to their association with Bessey's series of dicta intended to elucidate phylogenetic sequences. Another factor that first promoted interest in Bessey's system was his chart showing purported relationships of angiosperm orders.

The system of angiosperm classification of John Hutchinson (1884-1972) appeared in the first edition of his *Families of Flowering Plants*, London (1926-34). This underwent minor modification in a second edition, Oxford (1959). His classification of dicotyledons has been further revised and explained in his *Evolution and Phylogeny of Flowering Plants; Dicotyledons: Facts and Theory*, London (1969). His classification of monocotyledons with a fundamentally new interpretation of perianth characters is commonly viewed as an improvement over all earlier systems. His classification of dicotyledons based on two subclasses (when his terminology is made to conform to that in common use), one fundamentally woody and the other fundamentally herbaceous, has on the contrary met with much opposition. Hutchinson's position is explained in part by a list of dicta intended to elucidate phylogenetic progressions. He also provides charts showing purported relationships of angiosperm orders. He developed a key to the world's families of flowering plants and commenced a multi-volume *Genera of Flowering Plants*. A posthumous third edition of his *Families of Flowering Plants* (1973) maintains his system as formulated in 1969.

The system of angiosperm classification presented in the second volume (1964) of the twelfth edition of Engler's Syllabus der Pflanzenfamilien is here designated the Melchior system for the volume's editor, Hans Melchior (1894-). A number of collaborators have had a share in the system's classification of specific orders. The Melchior system, although it preserves many of the Engler and Prantl sequences, makes such sweeping concessions to Besseyan critics that it cannot be regarded as a neo-Englerian system. Although Melchior has abandoned many points in the Englerian system, critics contend he should also have abandoned others. But Melchior's conservatism seems less a defense of Engler's position than a rejection of alternatives deemed speculative. His review of current systems (pp. 14-26) emphasizes their diversity, and it bolsters his position in holding to old views in preference to transient new ones.

The system of angiosperm classification of Arthur J. Cronquist (1919-) is presented in his Evolution and Classification of Flowering Plants, Boston (1968). It adopts ten subclasses originally proposed by Armen Takhtajan (1910-) and parallels Takhtajan's system to a considerable extent. Cronquist optimistically suggests that taxonomists are now "all—or nearly all—Besseyans," and that the similarity of views "reflects the necessities of the present state of knowledge." The names of subclasses of the other systems considered in the present survey are descriptive in character, but those of Takhtajan and Cronquist are simply derivatives of ordinal names. Cronquist discusses a wide range of characters important to phylogenetic classification. He reviews the various taxa with reference to their display of such characters, and he provides synoptic keys designed to bring the taxa into an alignment appropriate to his system. He also provides charts showing purported relationships of orders within angiosperm subclasses.

O 3:1

TERMINOLOGY, NOMENCLATURE, SYNONYMY AND WORD FORMATION

The taxa recognized in the eight systems here considered are comparable, but terminological complications obscure the fact. Some regularization of categories is here adopted to facilitate comparisons. Endlicher's _sectio_ is here treated as a class; his _cohors_ is treated as a subclass; his _classis_ is treated as an order; and his _ordo_ is treated as a family. Bentham and Hooker's _subdivisio_ is here treated as a class. Where their _series_ and _cohors_ are used together they are treated as superorder and order respectively, but where their *series* is used without *cohorta* it is treated as an order. Their *ordo* is treated as a family. Eichler's *Abtheilung* is treated as a class; his *Klasse* is treated as a subclass; his *Reihe* is treated as a superorder; and his *Unterreihe* is treated as an order. Bessey's *phylum* is here treated as a division. Hutchinson's *phylum* is also here treated as a division; his *subphylum* is treated as a class; his *division* is treated as a subclass.

There are also numerous nomenclatural synonyms. Bessey's *Anthophyta* and Cronquist's *Magnoliophyta* are here treated as "Angiospermae." Endlicher's *Amphibrya*, Bentham and Hooker's and Hutchinson's *Monocotyledones*, Bessey's *Alternifoliae*, and Cronquist's *Liliatae* are treated as "Monocotyledoneae." Endlicher's *Acramphibrya*, Bentham and Hooker's and Hutchinson's *Dicotyledones*, Bessey's *Oppositifoliae*, and Cronquist's *Magnoliatae* are treated as "Dicotyledoneae." Except in the present terminological subsection and in the several glossaries the present survey employs a single terminology. These glossaries, which are indexed, should be consulted for further synonymy.

At the generic level three nomenclatural categories may be distinguished: (1) traditional, (2) descriptive, and (3) arbitrary. *Lilium* may be taken as an example of traditional generic nomenclature. It is the ancient latinized version of Greek *leirion*, "lily," undergoing a phonetic substitution of *l* for *r*. *Leirion* was later reborrowed as _lirion_, a form used in ancient botanical Latin. It provides the first part of the modern genus name _Liriodendron_, "lily tree." _Campanula_, "little bell," may be taken as an example of descriptive generic nomenclature. It is a medieval Latin name, but not an ancient Latin one. _Magnolia_, "plant of Magnolius" (for Pierre Magnol 1638-1715), may be taken as an example of arbitrary generic nomenclature. Although the name may relate to Magnol's introduction of American magnolia plants into the Montpellier botanical garden, it is arbitrary in that it lacks a traditional or descriptive basis.

At the family level two nomenclatural categories may be distinguished: (1) descriptive and (2) derivative. Descriptive names include: _Gramineae_, _Cruciferae_, _Leguminosae_, _Umbelliferae_, _Labiatae_, and _Compositae_. These are survivors of a group which has suffered much attrition. Among the lost names is, for example, _Asperifoliae_ for "Boraginaceae." These names are not based on generic names, and their termination in -_ae_ rather than in -_aceae_ emphasizes their distinction from derivative names. Derivative family names include _Liliaceae_, _Campanulaceae_, and _Magnoliaceae_. Note that the descriptive generic name _Campanula_ produces a derivative family name, not a descriptive one.

Ideally on Linnaean principles descriptive names are to be preferred over derivative names, but the practical difficulty of finding satisfactory ones has proved insurmountable. All new family names are therefore now required to be derivative. Some botanists discard all descriptive family names. For the names listed above, they substitute: Poaceae, Brassicaceae, Fabaceae, Apiaceae, Lamiaceae, and Asteraceae.

At the ordinal level the dichotomy between descriptive and derivative nomenclature reappears. Descriptive names include: Helobiae, Glumiflorae, Spathiflorae, Farinosae, Microspermae, Centrospermae, Parietales, Contortae, and Tubiflorae. These descriptive names are now largely given up in favor of such derivative names as: Alismatales, Graminales, Arales, Commelinales, Orchidales, Caryophyllales, Violales, Gentianales, and Polemoniales. A regular derivative name based on Umbelliferae would be Umbelliferales. The forms in use are irregular. They drop the -fer- and appear as Umbelliflorae and Umbellales.

At the subclass level the distinction between descriptive and derivative nomenclature persists. Here descriptive names, such as Archichlamydeae, Lignosae, and Strobiloideae, predominate. Endlicher, Bentham and Hooker, Eichler, Engler, Bessey, and Hutchinson each use a different set of descriptive subclass names. Melchior in most cases adopts those of Engler. Cronquist alone rejects descriptive subclass names and adopts derivative names, such as Alismatidae.

At the class level descriptive names, such as Monocotyledoneae, Alternifoliae, etc. predominate.

Cronquist alone adopts derivative names, such as _Liliatae_.

At the division level descriptive names, such as _Angiospermae_ and _Anthophyta_, again predominate. Cronquist alone prefers the derivative name _Magnoliophyta_.

0 4:1

EXPLANATION OF PLANT FORMULAE

Formulae have proved useful as a means of comparing and contrasting plant families and other taxa. Diverse systems of formulae have been devised. The system here presented adopts some of the conventions of earlier systems and introduces others. It is intended to be a flexible system susceptible of expansion to include any desired data. But it is also intended to avoid excess detail in any given formula. Thus the system is a specialized language for enumerating floral features and other plant characteristics. Translations are provided for the formulae given in connection with the families considered in the sections on specific orders. Formulae for individual specimens accompany the formulae for families.

The formula for the family Rhamnaceae is here given as an example. It is followed both by a translation and by an explanatory list of the symbols employed.

dPeri,Epi,(w.disk)G⁄2-4A4-5(opp.)C4-5K4-5/perf., polyg.-dioec.,drup.,caps.

Dicotyledonous plants with perigynous, or epigynous, flowers (having a disk). Gynoecium of 2 to 4 united carpels. Androecium of 4 to 5 stamens (opposite the petals). Corolla of 4 to 5 petals. Calyx of 4 to 5 sepals. Flowers perfect, or plants polygamodioecious. Fruit drupaceous, or a capsule.

d	Dicotyledonous plants	Peri	With perigynous flowers
,...,	Or	Epi	With epigynous flowers
(w.disk)	(With a disk)	G	Gynoecium of ... carpels
⁄	United	2-4	2 to 4
A	Androecium of ... stamens	(opp.)C	(Opposite the petals)
C	Corolla of ... petals	K	Calyx of ... sepals
/	.	perf.	Flowers perfect
polyg.-dioec.	Plants polygamodioecious	drup.	Fruit drupaceous
		caps.	Fruit a capsule

ANGIOSPERMAE AND COORDINATE GROUPS

The Angiospermae and Gymnospermae are not distinct from each other in all systems of classification. Endlicher includes the Gymnospermae as a subclass within the class Dicotyledoneae. Bentham and Hooker follow Endlicher in this. Eichler in his *Blüthendiagramme* splits off the Gymnospermae from the Dicotyledoneae and places both at class rank coordinate with the Monocotyledoneae. In a later system (1883) Eichler united the Monocotyledoneae and Dicotyledoneae at class rank into the division Angiospermae, coordinate to the division Gymnospermae. Engler, Hutchinson, and Melchior follow this phase of the later Eichler system. Bessey and Cronquist propose splitting the Gymnospermae into more than one division.

O 6:1

GLOSSARY FOR DIVISIONS, ETC.

Cormophyta (cf. Thallophyta). Regio (Endl.)

Phanerogamae (cf. Cryptogamae).

Divisio (B.&H.)

(Reich) (Eichl.)

Angiospermae (cf. Gymnospermae).

Divisio (Engl.)

Abteilung (Melch.)

Phylum (Hutch.)

Anthophyta (cf. Strobilophyta). Phylum (Bessey).

Magnoliophyta. Division (Cronq.)

O 7:1

ANGIOSPERM CLASSES

Endlicher recognizes the classes Monocotyledoneae and Dicotyledoneae in this sequence. Eichler and Engler also adopt this sequence. Bentham and Hooker, however, adopt the sequence: Dicotyledoneae, Monocotyledoneae. Hutchinson, Melchior, and Cronquist also adopt this sequence. In principle, Bessey agrees with the Bentham and Hooker sequence here. In early versions of his system, however, he followed Engler in regarding the two classes as independent, and in listing the class Monocotyledoneae first because it appeared to be morphologically simpler. Bessey persisted in following this sequence even after he explicitly argued for the derivation of the Monocotyledoneae from the Dicotyledoneae.

GLOSSARY FOR CLASSES, ETC.

Amphibrya, Acramphibrya. Sectiones in regio Cormophyta (Endl.)

Dicotyledones, Monocotyledones.

Subdivisiones in divisio Phanerogamae (B.&H.)

Subphyla in phylum Angiospermae (Hutch.)

Monocotyledoneae, Dicotyledoneae.

Abtheilungen in (Reich) Phanerogamae (Eichl.)

Classes in divisio Angiospermae (Engl.)

Klassen in Abteilung Angiospermae (Melch.)

Alternifoliae, Oppositifoliae. Classes in phylum Anthophyta (Bessey).

Magnoliatae, Liliatae. Classes in division Magnoliophyta (Cronq.) Cronquist has subsequently adopted the ending -opsida for class names.

MONOCOTYLEDON SUBCLASSES

Endlicher does not split the class Monocotyledoneae into subclasses. Bentham and Hooker, Eichler, Engler, and Melchior follow Endlicher in this. Bessey adopts two subclasses emphasizing ovary position. These subclasses may here be termed: Monocotostrobiloididae for hypogynous orders and Monocotocotyloididae for perigynous and epigynous orders. Hutchinson adopts three subclasses emphasizing perianth characteristics. These subclasses may here be termed: Calyciferidae in which at least the outer whorl of the perianth is calyx-like, Corolliferidae in which the entire perianth is corolloid, and Glumifloridae in which the flowers are associated with chaffy bracts. Cronquist adopts four subclasses. These are here given names derivative from ordinal names: Alismatidae, aquatic Calyciferidae; Commelinidae, Glumifloridae and terrestrial Calyciferidae; Aridae, spadiceous Corolliferidae; and Liliidae, aspadiceous Corolliferidae.

GLOSSARY FOR MONOCOTYLEDON SUBCLASSES, ETC.

Strobiloideae, Cotyloideae. Subclasses in class Alternifoliae (Bessey).

Calyciferae, Corolliferae, Glumiflorae. Divisions in subphylum Monocotyledones (Hutch.)

Alismatidae, Commelinidae, Arecidae, Liliidae. Subclasses in class Liliatae (Cronq.)

1 Oc:1

MONOCOTYLEDON ORDER GROUPS

Systems not explicitly employing subclasses within the Monocotyledoneae nevertheless follow recognizable patterns in their sequences of orders.

Endlicher follows the sequence: Graminales, Commelinales, Alismatales, Liliales, Iridales, Orchidales, Potamogetonales, Arales. This classification runs partly on a habit basis from grasses to trees, the palms. Palms not being within the flora here covered are not mentioned in the sequence. The sequence can be broken into four order groups analogous to Cronquist's subclasses: (1) Graminales, Commelinales, (2) Alismatales, (3) Liliales, Iridales, Orchidales, (4)

Potamogetonales, Arales. Ordinal names corresponding to subclass names are here distinguished from the others. The Liliales, Iridales, and Orchidales are associated by corolliform perianth.

Bentham and Hooker employ a sequence diverging widely from that of Endlicher. It runs: Orchidales, Iridales, Liliales (incl. Commelinaceae), Juncales, Arales, Alismatales, Graminales. This may be interpreted as an arrangement which gives the Liliales a pivotal position. The first two orders constitute one line of divergence based on ovary position inferior. The third order has superior ovary and a petaloid perianth. The fourth also has superior ovary, but it lacks the petaloid character in its perianth. The final orders are based on: perianth lacking, apocarpy, and glumaceous inflorescence. The sequence can be broken into four order groups analogous to Cronquist's subclasses: (1) Orchidales, Iridales, Liliales (incl. Commelinaceae), Juncales, (2) Arales, (3) Alismatales, (4) Graminales.

The orders of Eichler's monocotyledons run: Alismatales, Arales, Graminales, Commelinales, Liliales, Orchidales. The sequence contrasts the free carpellate or functionally unicarpellate structure of the first three orders with the united and functionally tricarpellate structure typical of the last three. The sequence can be broken into five order groups analogous to Cronquist's subclasses: (1) Alismatales, (2) Arales, (3) Graminales, (4) Commelinales, (5) Liliales, Orchidales.

Engler modifies Eichler's sequence of monocot orders. He splits off the family Typhaceae from the Arales, raises it to ordinal rank, and gives it first place in his system. He also transposes the Arales and Graminales. Both of these changes result in part from his laying stress on the aralian spadix--first to exclude the Typhaceae (incl. _Sparganium_) from the Arales, and secondly to separate the Arales from the rest of the Eichlerian orders with functionally unicarpellate pistils. Thus he arrives at the sequence: Typhales, Alismatales, Graminales, Arales, Commelinales, Liliales, Orchidales. The sequence can be broken into six order groups analogous to Cronquist's subclasses: (1) _Typhales_, (2) _Alismatales_, (3) _Graminales_, (4) _Arales_, (5) _Commelinales_, (6) _Liliales_, Orchidales.

Bessey's Monocotostrobiloididae orders have the sequence: Alismatales, Liliales (incl. Commelinaceae), Arales, Graminales. His diagrams derive the Alismatales from the ranalian dicotyledons, and the Liliales from the Alismatales. These orders may be interpreted as analogous to order groups: (A1) _Alismatales_, (A2) _Liliales_, (A3) _Arales_, (A4) _Graminales_. To these may be added from his Monocotocotyloididae: (B1) _Iridales_, Orchidales. The Iridales he derives from the Liliales, and the Orchidales from the Iridales. The system lays special stress on the many carpels of certain alismatalian groups and on epigyny.

Hutchinson's orders in the Calyciferidae, the first of his three monocot subclasses, have the sequence: Alismatales, Potamogetonales, Commelinales. His diagrams derive the Alismatales from a predominantly herbaceous ranalian group of dicotyledons, and they derive the Potamogetonales and Commelinales separately from the Alismatales. The sequence may be interpreted as two order groups: (A1) Alismatales, Potamogetonales, (A2) Commelinales. His orders in the Corolliferidae, the second of his monocot subclasses, have the sequence: Liliales, Arales, Typhales, Amaryllidales, Iridales, Orchidales. His diagrams derive the Liliales from the Calyciferidae; and from the Liliales he derives the Arales, Typhales, and Orchidales, each group being independent in its development. He also derives from the Liliales a line which gives rise to the Amaryllidales and Iridales. This sequence may be interpreted as five order groups: (B1) Liliales, (B2) Arales, (B3) Typhales, (B4) Amaryllidales, Iridales, (B5) Orchidales. The corolliform character is evident for most of these orders, but the Arales and Typhales require special explanation. Hutchinson regards the Arales as reduced from the Aspidistreae in the Liliaceae, a tribe with dense floral spikes and wide leaves. More data on this theory are given in the section on the Arales. Hutchinson regards the Typhales as reduced from the ancestral stock of the Xanthorrhoeaceae, a lilialian family with dense floral spikes. More data on this theory are given in the section on the Typhales. Hutchinson's orders in the Glumifloridae, the third of his monocot subclasses, have the sequence: Juncales, Cyperales, Graminales. His diagrams derive the Juncales from the Liliales, and they derive the Cyperales and Graminales separately from the Juncales. Progressive reduction is the basic tendency for this subclass. The sequence may be interpreted as a single order group: (Cl) Juncales, Cyperales, <u>Graminales</u>.

Melchior's sequence of monocot orders runs: Alismatales, Liliales, Juncales, Commelinales, Graminales, Arales, Typhales, Cyperales, Orchidales. Melchior admits the Alismatales to first position among monocot orders, and from taxa in or near the family Alismataceae he derives the Liliales. He accepts relationships between the Liliales, Juncales, Commelinales, and Arales. These relationships are not made explicit, but they appear to derive from primitive lilialian taxa. On the other hand he considers the derivation and relationships of the Typhales, Cyperales, and Orchidales to be unknown. The sequence may be interpreted as seven order groups: (1) <u>Alismatales</u>, (2) <u>Liliales</u>, Juncales, (3) <u>Commelinales</u>, Graminales, (4) <u>Arales</u>, (5) <u>Typhales</u>, (6) <u>Cyperales</u>, (7) <u>Orchidales</u>.

Cronquist's subclass Alismatidae corresponds to a predominantly aquatic portion of the Hutchinsonian subclass Calyciferidae. The subclass Alismatidae contains the orders: (Al) <u>Alismatales</u>, Potamogetonales. Cronquist suggests the Alismatales derive from forms close to the Nymphaeales. He also holds that they stand near but not on the base of a line leading to the other subclasses of monocotyledons. Cronquist's subclass Commelinidae corresponds to a predominantly terrestrial portion of the Hutchinsonian subclass Calyciferidae plus the subclass Glumifloridae. The subclass Commelinidae has the following sequence of orders: (Bl) Commelinales, Juncales, Graminales, Typhales. Cronquist diagrams the Commelinidae as derived from a line originating near the Alismatidae and passing near the Aridae before diverging into the Commelinidae and Liliidae. He diagrams two offshoots from the Commelinales. One of them by branching leads to both the Juncales and Graminales. The other leads to the Typhales. Cronquist's subclass Aridae corresponds to the spadiciferous portion of the Hutchinsonian subclass Corolliferidae. The subclass Aridae contains the order: (Cl) Arales. It also includes a number of tropical taxa, among them the palms, which are not here considered. Cronquist's subclass Liliidae corresponds to the non-spadiciferous portion of the Hutchinsonian subclass Corolliferidae. The subclass Liliidae contains the orders (Dl) Liliales, Orchidales; and Cronquist explicitly derives the Orchidales from the Liliales. In the systems of Bentham and Hooker, Bessey, and Hutchinson the Liliales appear the pivotal monocot order, the relationships of which determine the general character of the classification; but in Cronquist's system it is the Commelinales which have this pivotal position.

TYPHALES

The families of the Typhales here considered are the Typhaceae and Sparganiaceae. The family Typhaceae contains the single genus *Typha*, cat-tail. The family Sparganiaceae contains the single genus *Sparganium*, bur-reed.

Among the characters that have been used in classification as part of synopses to distinguish the order are some that entail difficulties. The character of a single carpel in pistillate flowers of the Typhales meets with exceptions, but it may be restated as the character of a functionally unicarpellate pistil. Again a lack of perianth in the Typhales has been used as a distinguishing character, but the bractish parts of the *Sparganium* flowers are now usually interpreted as perianth, and the hairs surrounding the *Typha* flowers seem homologous.

Typha is a pre-Linnaean Latin word based on an ancient Greek name for the cat-tail. *Sparganium* is derived through ancient Latin from a Greek name for the bur-reed. The word is interpreted as signifying "swaddling-band" in allusion to the long narrow leaves of the plant.

TYPHACEAE

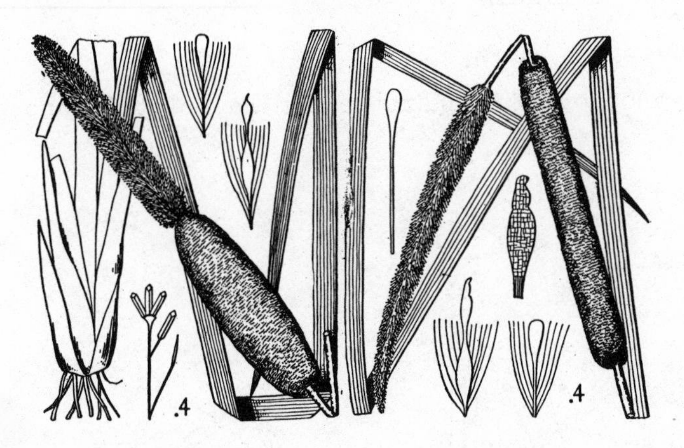

Typha latifolia	Typha angustifolia
-	-
cattail	cattail

TYPHACEAE

øHypoGl,1̸2,gynophore(f)A2-7androphore(m)Pn-few(brist.)/
monoec.nutlet paludal/vs.PO

Monocotyledonous plants with hypogynous flowers. Gynoecium of a single carpel, or of 2 united carpels. The ovary on a gynophore (in the case of pistillate flowers). Androecium of 2 to 7 stamens. The stamens on an androphore (in the case of staminate flowers). Perianth of numerous to few tepals (bristles). Plants monoecious.

Fruit a nutlet. Plants paludal. By alternative interpretation: Perianth lacking.

Typha sp.

∅HypoG1gynophore(f)A3androphore(m)Pn(brist.)/monoec.

SPARGANIACEAE

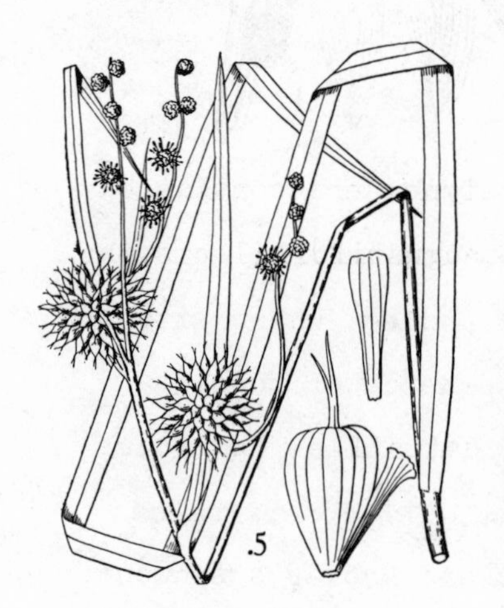

Sparganium eurycarpum

-

bur-reed

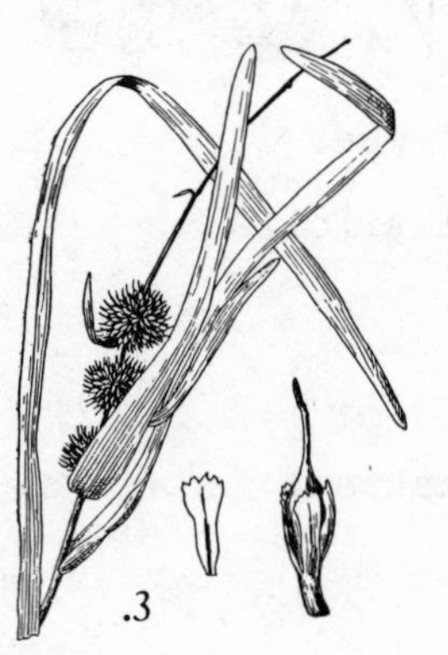

Sparganium americanum

-

bur-reed

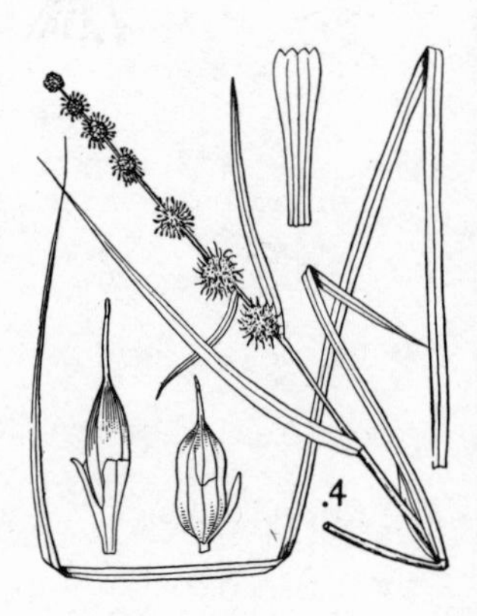

Sparganium angustifolium

-

bur-reed

SPARGANIACEAE

ØHypoG1,Ø2,A3-6P3-6(scales)/monoec.ach.aquat./vs.PO

Monocotyledonous plants with hypogynous flowers. Gynoecium of a single carpel, or of 2 united carpels. Androecium of 3 to 6 stamens. Perianth of 3 to 6 tepals (scales). Plants monoecious. Fruit an achene. Plants aquatic. By alternative interpretation: Perianth lacking.

Sparganium sp.

ØHypoG1A3P3(scales)/monoec.

1 3:1

Of the eight classification systems here considered four do not recognize the Typhales. Endlicher places the Typhaceae and Sparganiaceae in the Arales, which with the Palmales are near the end of his sequence of monocot orders. Bentham and Hooker and Eichler follow Endlicher in placing these taxa in the Arales. The order Arales comes toward the middle of Bentham and Hooker's sequence of monocot orders, and it follows the Liliales and Juncales. These orders are in Bentham and Hooker's own terminology: Coronarieae, Calycinae, and Nudiflorae. It is then on the basis of a lack of perianth that they contrast the Arales with the Liliales and Juncales. Eichler classifies the Arales in the first of his two basic groups of monocot orders—that having functionally unicarpellate pistils. The Arales follow the Alismatales in this group.

Engler splits off the Typhaceae and Sparganiaceae from the Arales, raises them to ordinal rank as the Typhales, and on the basis of floral simplicity puts this order at the beginning of his monocot classification. Bessey accepts Engler's separation of the Typhaceae and Sparganiaceae from the Arales. He rejects, however, the concept of an order Typhales, and he alone places the Typhaceae and Sparganiaceae within the Alismatales. This order stands at the beginning of his monocot classification.

Hutchinson, Melchior, and Cronquist all accept the Typhales as an advanced order. Hutchinson places the order in the Corolliferidae, the second of his three monocot subclasses and after the Liliales, from which order he holds the Typhales to be derived. More specifically, Hutchinson accounts for the Typhales as related to and reduced from the lilialian ancestral stock of the agavalian family Xanthorrhoeaceae, "wherein the inflorescence tends toward the densely spicate type." Compare Hutchinson's sketches for the Typhaceae (Families of Flowering Plants, ed. 2, v. 2, p. 638) with those for the Xanthorrhoeaceae (op. cit., p. 661). Melchior places the Typhales toward the end of his monocot classification as an order of unknown derivation. Cronquist places the Typhales in the Commelinidae, the second of his four monocot subclasses and after the Commelinales from which he holds the Typhales to be derived. His argument for this derivation is not based on specific analogy such as the citation of the Xanthorrhoeaceae by Hutchinson. Instead he cites stomatal organization, vessel distribution, wind pollination, etc. as characters providing a smoother transition from the Commelinales to the Typhales than from the Liliales to the Typhales.

There is then consensus that the order Typhales is advanced rather than primitive (Engler dissenting), but opinions differ as to whether it has alismatalian, aralian, commelinalian, lilialian, or unknown origin.

There is agreement that *Typha* and *Sparganium* are related. Endlicher, Bentham and Hooker, and Eichler place both genera in the Typhaceae. Cronquist also inclines to this position, but he retains the Sparganiaceae as a distinct family out of deference to the Englerian tradition which segregates *Typha* and *Sparganium* into separate families. In the systems which recognize both families, those of Engler and Bessey give first place to the Typhaceae. Hutchinson, Melchior, and Cronquist, however, agree in placing the Sparganiaceae first. There is then a modern consensus that the Typhaceae are more advanced than the Sparganiaceae. In support of this view reference can be made to the bractish tepals of the *Sparganium* flower as being nearer to a normal perianth than are the hairs or bristles of *Typha*. Also the inflorescences of *Sparganium* are nearer to simple condensed inflorescences than are the cylindrical inflorescences of Typha.

1 4:1

GLOSSARY FOR TYPHALES

Typhales. Order in class Monocotyledoneae.
  Pandanales. Series in classis Monocotyledoneae (Engl.)
  Typhales. Order in division Corolliferae (Hutch.)
  Pandanales. Reihe in Klasse Monocotyledoneae (Melch.)
  Typhales. Order in subclass Commelinidae (Cronq.)
Typhaceae. Family in order Typhales.
  Typhaceae. Ordo in classis Spadiciflorae (Endl.)
  Typhaceae. Ordo in series Nudiflorae (B.&H.)
  Typhaceae. Familie in Unterreihe Spadiciflorae (Eichl.)
  Typhaceae. Familia in series Pandanales (Engl.)
  Typhaceae. Family in order Alismatales (Bessey).
  Typhaceae. Family in order Typhales (Hutch.), (Cronq.)
  Typhaceae. Familie in Reihe Pandanales (Melch.)
Sparganiaceae. Family in order Typhales.
  Sparganiaceae. Familia in series Pandanales (Engl.)
  Sparganiaceae. Family in order Alismatales (Bessey).
  Sparganiaceae. Family in order Typhales (Hutch.), (Cronq.)
  Sparganiaceae. Familie in Reihe Pandanales (Melch.)

ALISMATALES

The Alismatales include several families, but only the Alismataceae and Potamogetonaceae are considered here. These two families consist of aquatic or marsh herbs.

The character: "perianth present" is sometimes used in synoptic keys to separate the order Alismatales from the Typhales. This requires that the flower of Sparganium not be interpreted as having bractish tepals and also that the flower of Potamogeton be interpreted as having tepals rather than merely tepal-like stamen appendages. One interpretation of the flower of Potamogeton takes it to be, in fact, a pseudanthium of four staminate flowers, each with a bract, surrounding a single pistillate flower.

The flowers of the Alismataceae have numerous free carpels. This feature has been interpreted as suggesting that the family has floral structure derived from that of the ranalian dicots and that the flowers of the other monocotyledons may be viewed as derived from an alismatacean type.

Potamogeton is the ancient Greek name for the pondweed. It is derived from potamos, "river," and geiton, "borderer." the name thus signifies literally: "plant which lives in the margin of a river." Alisma is the ancient Greek name for the water plantain. The literal meaning of this name is obscure.

2 2:1

POTAMOGETONACEAE

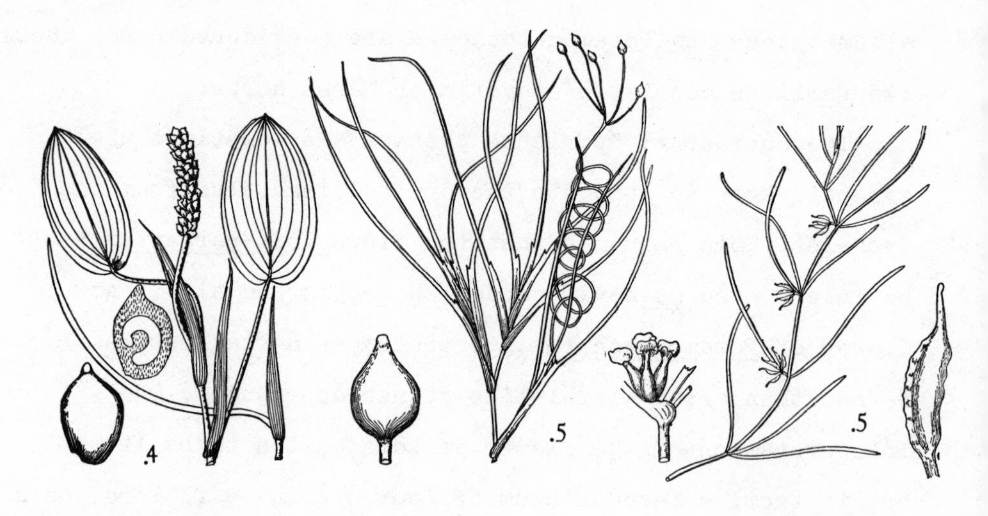

Potamogeton natans	Ruppia oxidentalis	Zannichellia palustris
- pondweed	- ditch grass	horned pondweed

POTAMOGETONACEAE

øHypoG4-1A(epitep.opp.)4,A2,P4,0/nutlet,drup.aquat./vs. A4,2,PO

Monocotyledonous plants with hypogynous flowers. Gynoecium

of 4 to 1 carpels. Androecium (epitepalous and opposite the tepals) of 4 stamens, or androecium (hypogynous) of 2 stamens. Perianth of 4 tepals, or none. Fruit a nutlet, or drupaceous. Plants aquatic. By alternative interpretation: Androecium of 4 stamens, or 2. Perianth lacking.

Potamogeton sp.

đHypoG4A(epitep.opp.)4P4/

ALISMATACEAE

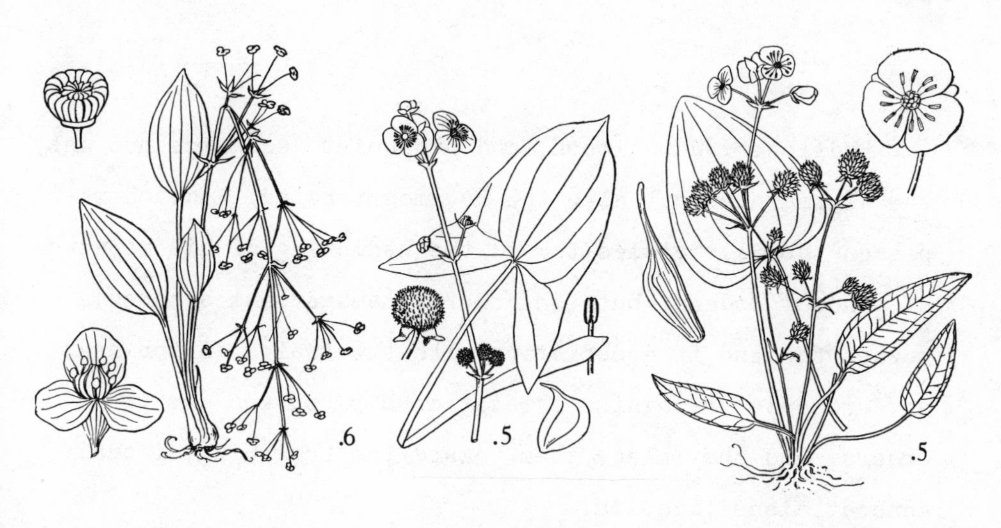

Alisma subcordatum	Sagittaria latifolia	Echinodorus cordifolius
-	-	-
water plantain	arrowhead	burhead

ALISMATALES 2 2:3

ALISMATACEAE

¢HypoG6-nA6-nC3K3/perf.,monoec.,dioec.,ach.aq.,paludal

Monocotyledonous plants with hypogynous flowers. Gynoecium of 6 to many carpels. Androecium of 6 to many stamens. Corolla of 3 petals. Calyx of 3 sepals. Flowers perfect, or unisexual and the plants monoecious, or dioecious. Fruit an achene. Plants aquatic, or paludal.

Alisma sp.

¢HypoGnA6C3K3/

Sagitaria sp.

¢HypoGnAnC3K3/

2 3:1

Endlicher, Hutchinson, and Cronquist recognize not only the Alismatales but also the Potamogetonales. Endlicher places the Alismatales toward the beginning of his sequence of monocot orders, but the order Potamogetonales comes toward the end in association with the Arales. Hutchinson and Cronquist recognize a relationship between the two orders, and they place them toward the beginning of their monocot classification.

Bentham and Hooker, Eichler, Engler, Bessey, and Melchior include the Potamogetonaceae within their Alismatales. Bentham and Hooker place the Alismatales near the end of their monocot order sequence. Eichler, on the other hand, places the Alismatales in first position among his monocot orders.

Engler places the Typhales ahead of the Alismatales at the beginning of his sequence of monocot orders. Bessey not only places the order Alismatales in first position, but he derives the other monocot orders from it. Later systems adopt the Besseyan position on this, but with qualifications.

There is then a modern consensus that the Alismataceae are related to the Potamogetonaceae. There is also a modern consensus that the order Alismatales includes some primitive taxa.

The three systems which recognize the two orders all accord first place to the Alismatales. Two of the five systems which recognize only the one order, those of Eichler and Engler, accord first place to the Potamogetonaceae. The other three, those of Bentham and Hooker, Bessey, and Melchior, give first place to the Alismataceae. The one interpretation accepts the apparent simplicity of the Potamogetonaceae as primitive. The other interpretation accepts the same simplicity as evidence of reduction. Bessey, Hutchinson, and Cronquist see in the Alismatales primitive conditions which connect the monocotyledons directly or indirectly to the ranalian dicotyledons.

GLOSSARY FOR ALISMATALES

Potamogetonales. Order in class Monocotyledoneae.

Fluviales. Classis in sectio Amphibrya (Endl.)

Potamogetonales. Order in division Calyciferae (Hutch.)

Najadales. Order in subclass Alismatidae (Cronq.)

Alismatales. Order in class Monocotyledoneae.

Helobiae. Classis in sectio Amphibrya (Endl.)

Apocarpae. Series in subdivisio Monocotyledones (B.&H.)

Helobiae. Unterreihe in Abtheilung Monocotyledoneae (Eichl.)

Helobiae. Series in classis Monocotyledoneae (Engl.)

Alismatales. Order in subclass Strobiloideae (Bessey).

Alismatales. Order in division Calyciferae (Hutch.)

Helobiae. Reihe in Klasse Monocotyledoneae (Melch.)

Alismatales. Order in subclass Alismatidae (Cronq.)

Potamogetonaceae. Family in order Alismatales.

Najadeae. Ordo in classis Fluviales (Endl.)

Naiadaceae. Ordo in series Apocarpae (B.&H.)

Najadaceae. Familie in Unterreihe Helobiae (Eichl.)

Potamogetonaceae. Familia in series Helobiae (Engl.)

Potamogetonaceae. Family in order Alismatales (Bessey).

Potamogetonaceae. Family in order Potamogetonales (Hutch.)

Potamogetonaceae. Familie in Reihe Helobiae (Melch.)

Potamogetonaceae. Family in order Najadales (Cronq.)

Alismataceae. Family in order Alismatales.

Alismaceae. Ordo in classis Helobiae (Endl.)

Alismaceae. Ordo in series Apocarpae (B.&H.)

Alismaceae. Familie in Unterreihe Helobiae (Eichl.)

Alismataceae. Familia in series Helobiae (Engl.)

Alismataceae. Family in order Alismatales (Bessey), (Hutch.), (Cronq.)

Alismataceae. Familie in Reihe Helobiae (Melch.)

Typhaceae. See 1 4.

Typhaceae. Family in order Alismatales (Bessey).

Sparganiaceae. See 1 4.

Sparganiaceae. Family in order Alismatales (Bessey).

3 1:1

GRAMINALES

The Graminales include the families Cyperaceae and Gramineae, the sedges and grasses. Both contain numerous important genera which are usually easily recognized as belonging to the group. Various characters can be used for identification to order. Especially distinctive are the minute flowers lacking tripartite perianth and grouped in spikes or spikelets, and the inflorescence subtended by chaffy bracts.

The differentiation between the Cyperaceae and Gramineae is usually easy, although both groups are variable and may have similar general aspect. Indeed the fact that the groups are so different has led some to place them remote from one another in classification systems. The stems of the Cyperaceae typically are solid and triangular in cross section, the leaf sheaths surrounding them closed and the leaves three-ranked, in contrast to the Gramineae with typically hollow and cylindrical stems, the leaf sheaths surrounding them open and the leaves two-ranked.

Gramineae is a descriptive family name formed as a feminine plural from Latin *gramineus* "grassy." It signifies literally: "Family of grassy plants." *Cyperus* is the ancient Greek name for sedge.

3 2:1

GRAMINEAE

ɇHypoG⨯3-1,q,L1A3P2(lodic.),0,B2(pal.+lem.)/perf.,monoec., dioec.,grain glumes/vs.P2(lodic.(C)),0,+⨯2(pal.(K))B1(lem.)/ vs.B2(lodic.),0,+2(pal.+lem.)

Monocotyledonous plants with hypogynous flowers. Gynoecium of 3 to 1 carpels, united when more than 1, or with carpellary number not evident. The ovary containing a single locule. Androecium of 3 stamens. Perianth of 2 members (lodicules), or none. Bracts 2 (palea and

GRAMINEAE

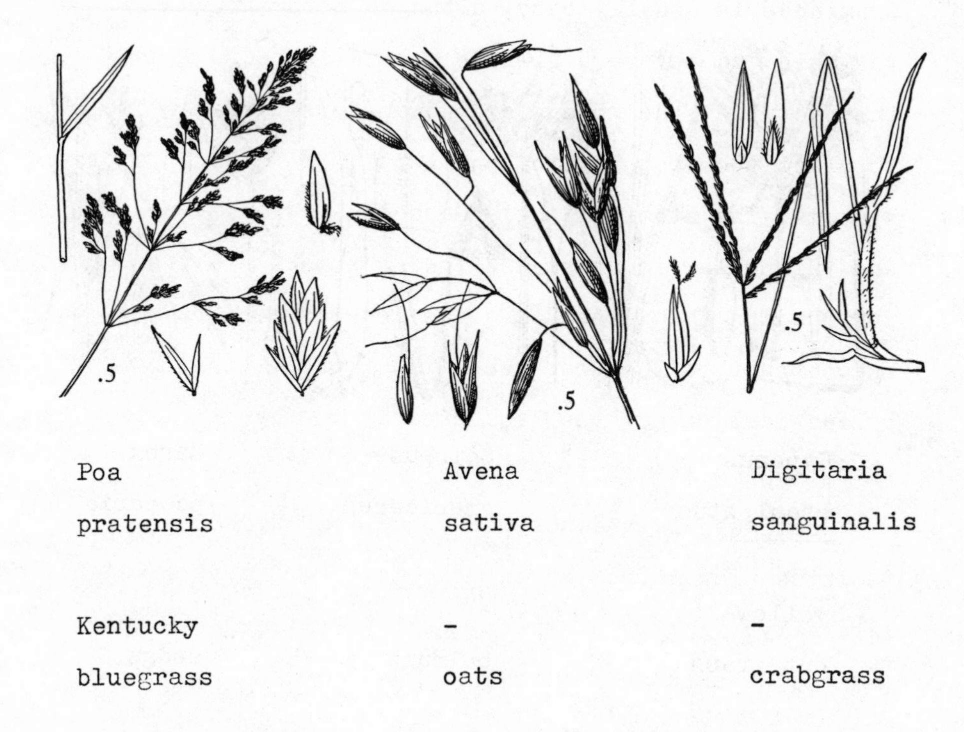

Poa pratensis	Avena sativa	Digitaria sanguinalis
Kentucky bluegrass	- oats	- crabgrass

lemma). Flowers perfect, or unisexual and the plants monoecious or dioecious. Fruit a grain (caryopsis). Glumes present. By alternative interpretation: Perianth of 2 tepals (lodicules (of corolloid origin)), or none, plus 2 united tepals (palea (of sepaloid origin)). Bract 1 (lemma). By alternative interpretation: Bracts 2 (lodicules), or none, plus 2 (palea and lemma).

Poa sp.

¢HypoG2A3P2(lodic.)B2(pal.+lem.)/

CYPERACEAE

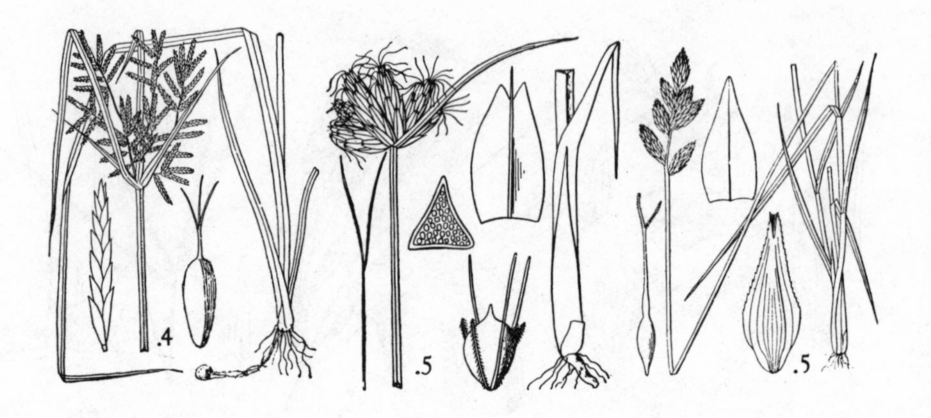

Cyperus	Scirpus	Carex
esculentus	americanus	scoparia
yellow	-	-
nut-grass	bulrush	sedge

CYPERACEAE

øHypoG/3-2L1A3Pn-O(brist.,scales)Bl,2/perf.,monoec., ach.,utr.,glumes.

Monocotyledonous plants with hypogynous flowers. Gynoecium of 3 to 2 united carpels, the ovary with a single locule. Androecium of 3 stamens. Perianth of numerous members to none (bristles or scales). Bracts 1, or 2. Flowers perfect, or unisexual and the plants monoecious. Fruit an achene, or utricle. Glumes present.

Cyperus sp.

d̸HypoGi̸2L1A3B1/perf.

Scirpus sp.

d̸HypoGi̸3L1A3P6(brist.)B1/perf.

Carex sp.

d̸HypoGi̸L1A3B2(perigynium+bract(f)),1(m)/monoec.

Bracts 2 (a perigynium and a bract (in pistillate flowers)), or a single bract (in staminate flowers).

3 3:1

Six of the eight systems here considered include the Cyperaceae within the Graminales. Those of Hutchinson and Melchior, however, split off the family Cyperaceae and raise it to ordinal rank. Hutchinson retains the two orders in juxtaposition, but Melchior places the Cyperales far from the Graminales. This separation is based on an interpretation of the cyperalian flower as a pseudanthium (cf. Syllabus der Pflanzenfamilien, ed. 12, v. 2, p. 604, fig.)

Endlicher places the Graminales at the beginning of his monocot order sequence in association with the Commelinales. Bentham and Hooker, however, exclude the Graminales from any connection with their Commelinales. The order Graminales stands isolated at the end of their monocot classification. Eichler employs a distinction between orders with functionally unicarpellate pistils and orders with functionally tricarpellate ones. Eichler's order Graminales stands isolated, last in the first group. This isolation of the Graminales is maintained in the Englerian system, although the Eichler group is modified in arrangement by Engler. Bessey places the Graminales last in the Monocotostrobiloididae, the first of his two monocot subclasses. He proposes as a phylogenetic sequence: Alismatales, Liliales, Graminales. Hutchinson isolates the Graminales together with the Juncales and Cyperales in the Glumifloridae, the last of his three monocot subclasses. Like Bentham and Hooker he makes the Graminales the last monocot order. He proposes as a phylogenetic sequence: Alismatales, Liliales, Juncales, with the Cyperales and Graminales separately derived from the Juncales. Melchior places the Graminales toward the middle of his sequence of monocot orders, directly after the Commelinales. He points to the Restionaceae as a connecting link between the two orders (op. cit. p. 557; p. 558, fig.) His Cyperales are in isolated position toward the end of his sequence of monocot orders, and he holds their origin to be unknown. Cronquist includes the Graminales in the Commelinidae, the second of his four monocot subclasses. He proposes a phylogenetic line from the Commelinales which divides to lead to both the Juncales and Graminales. Although both Melchior and Cronquist derive the Graminales from the Commelinales, their positions are quite different. Melchior makes the derivation through the Restionaceae, and he excludes the Cyperaceae from the Graminales. Cronquist denies the Restionaceae any place on the line leading to the Graminales, and he includes the Cyperaceae in the order.

Endlicher places the Gramineae ahead of the Cyperaceae. Bentham and Hooker adopt the reverse sequence. Engler follows the sequence of Endlicher, but Eichler, Bessey, and Cronquist follow that of Bentham and Hooker. Hutchinson places his Cyperales ahead of the Graminales.

There is then consensus (Melchior dissenting) that the Cyperaceae and Gramineae are closely enough related to be placed in the same order, or at least in adjacent orders (Hutchinson). There is also a preponderance of opinion that the Gramineae are more advanced than the Cyperaceae. In support of this view advanced features of the Gramineae may be cited, such as the caryopsis fruit, hollow stems, and two-ranked leaves. Intercalary meristem may be noted as an adaptation to grazing.

3 4:1

GLOSSARY FOR GRAMINALES

Graminales. Order in class Monocotyledoneae.

Glumaceae. Classis in sectio Amphibrya (Endl.)

Glumaceae. Series in subdivisio Monocotyledones (B.&H.)

Glumaceae. Unterreihe in Abtheilung Monocotyledoneae (Eichl.)

Glumiflorae. Series in classis Monocotyledoneae (Engl.)

Graminales. Order in subclass Strobiloideae (Bessey).

Graminales. Order in division Glumiflorae (Hutch.)

Graminales. Reihe in Klasse Monocotyledoneae (Melch.)

Cyperales. Order in subclass Commelinidae (Cronq.)

Cyperales. Order in class Monocotyledoneae.

Cyperales. Order in division Glumiflorae (Hutch.)

Cyperales. Reihe in Klasse Monocotyledoneae (Melch.)

Gramineae. Family in order Graminales.

Gramineae. Ordo in classis Glumaceae (Endl.)

Gramineae. Ordo in series Glumaceae (B.&H.)

Gramina. Familie in Unterreihe Glumaceae (Eichl.)

Gramineae. Familia in series Glumiflorae (Engl.)

Poaceae. Family in order Graminales (Bessey).

Gramineae. Family in order Graminales (Hutch.)

Gramineae. Familie in Reihe Graminales (Melch.)

Gramineae. Family in order Cyperales (Cronq.)

Cyperaceae. Family in order Graminales.

Cyperaceae. Ordo in classis Glumaceae (Endl.)

Cyperaceae. Ordo in series Glumaceae (B.&H.)

Cyperaceae. Familie in Unterreihe Glumaceae (Eichl.)

Cyperaceae. Familia in series Glumiflorae (Engl.)

Cyperaceae. Family in order Graminales (Bessey).

Cyperaceae. Family in order Cyperales (Hutch.)

Cyperaceae. Familie in Reihe Cyperales (Melch.)

Cyperaceae. Family in order Cyperales (sensu Graminales)
(Cronq.)

4 1:1

ARALES

The order Arales includes the Araceae and the Lemnaceae. The Lemnaceae are reduced aquatics, and their putative relationship to the Araceae rests in part on technical embryological arguments. Only the Araceae are here considered. These have a distinctive fleshy flowering spike known as a spadix. Among the genera of the family which occur naturally in our area are *Acorus* and *Arisaema*. *Arum*, *Monstera*, and *Anthurium* are among the several genera of the family with species cultivated as ornamentals.

Arum is an ancient Latinized version of the Greek name for certain araceous plants.

4 2:1

ARACEAE

ȼHypoG⁄1-3A4-6P4-6,0/perf.,monoec.,dioec.,ber.spad.w.
spathac.B

ARACEAE

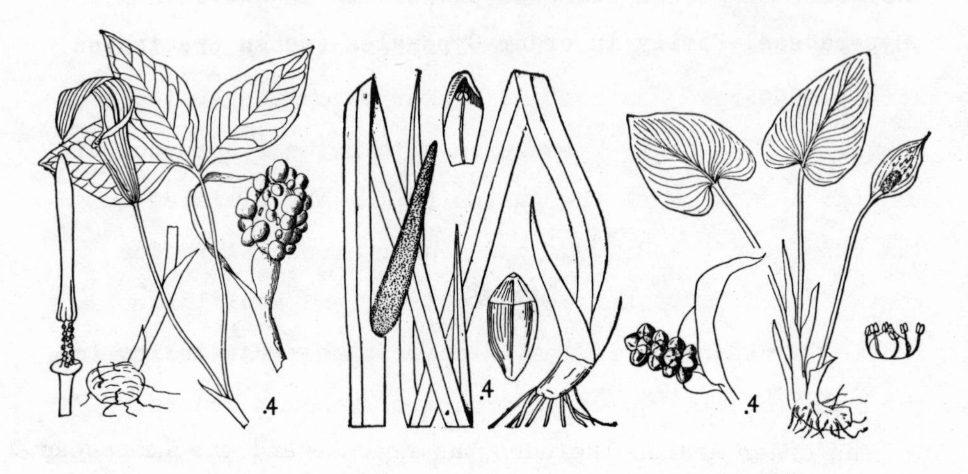

Arisaema triphyllum	Acorus calamus	Calla palustris
Jack-in-the pulpit	sweetflag	water arum

Monocotyledonous plants with hypogynous flowers. Gynoecium of 1 to 3 carpels, united when more than 1. Androecium of 4 to 6 stamens. Perianth of 4 to 6 tepals, or none. Flowers perfect, or unisexual and the plants monoecious or dioecious. Fruit a berry. Inflorescence a spadix with a spathaceous bract.

Acorus calamus

d̸HypoG1̸3A6P6/perf.

Endlicher classifies the Arales next to the Palmales. The order Palmales, which he places at the end of his monocot classification, is not represented in the flora here considered. Bentham and Hooker place the Arales in a more or less central position within their sequence of monocot orders. Here the Arales follow the Liliales and Juncales, and the Bentham and Hooker nomenclature for these orders: Coronarieae, Calycinae, and Nudiflorae implies a reduction series. Eichler places the Arales in the first of his two basic groups of monocot orders. These orders typically have the carpels free or the pistil functionally unicarpellate. In this group the Arales are second only to the Alismatales. On the basis of their spadix, Engler separates the Arales from the rest of Eichler's first group. This leaves the order in the last position within that group. Bessey places the Arales after the Alismatales and Liliales in the Monocotostrobiloididae, the first of his two monocot subclasses; and he diagrams as a phylogenetic sequence: Alismatales, Liliales, Arales. Hutchinson places the Arales second only to the Liliales in the Corolliferidae, the second of his three monocot subclasses. Like Bessey he derives the Arales from the Liliales. He specifies forms such as _Lysichitum_ (cf. _Families of Flowering Plants_, ed. 2, v. 2, p. 628, fig.) to be primitive in the Arales, and he suggests these forms point to aralian ancestry in the

Aspidistrae (op. cit. p. 629; p. 603, fig.) of the Liliaceae. The tribe Aspidistrae has dense floral spikes and wide leaves. Melchior begins his sequence of monocot orders with the Alismatales followed by the Liliales. Then come orders he derives from the Liliales, among which is the order Arales. As a phylogenetic sequence Melchior adopts Bessey's proposal: Alismatales, Liliales, Arales. Cronquist derives the Arales directly from a line arising near the Alismatidae. He accepts the aralian order as remote from other monocots and on it therefore bases the Aridae, the third of his four monocot subclasses.

The various systems differ in their interpretation of the relationship, if any, of the Typhaceae and Sparganiaceae to the Arales. For details on this see the discussion in the section on the Typhales. The systems prior to Engler's included *Typha* and *Sparganium* in the Arales. Engler on the basis of their lack of spadix, excluded these taxa from the aralian order and proposed the order Typhales. The subsequent systems have concurred in this separation.

GLOSSARY FOR ARALES

Arales. Order in class Monocotyledoneae.

Spadiciflorae. Classis in sectio Amphibrya (Endl.)

Nudiflorae. Series in subdivisio Monocotyledones (B.&H.)

Spadiciflorae. Unterreihe in Abtheilung Monocotyledoneae (Eichl.)

Spathiflorae. Series in classis Monocotyledoneae (Engl.)

Arales. Order in subclass Strobiloideae (Bessey).

Arales. Order in division Corolliferae (Hutch.)

Spathiflorae. Reihe in Klasse Monocotyledoneae (Melch.)

Arales. Order in subclass Arecidae (Cronq.)

Araceae. Family in order Arales.

Aroideae. Ordo in classis Spadiciflorae (Endl.)

Aroideae. Ordo in series Nudiflorae (B.&H.)

Araceae. Familie in Unterreihe Spadiciflorae (Eichl.)

Araceae. Familia in series Spathiflorae (Engl.)

Araceae. Family in order Arales (Bessey), (Hutch.), (Cronq.)

Araceae. Familie in Reihe Spathiflorae (Melch.)

Typhaceae. See 1 4.

Typhaceae. Ordo in classis Spadiciflorae (Endl.)

Typhaceae. Ordo in series Nudiflorae (B.&H.)

Typhaceae. Familie in Unterreihe Spadiciflorae (Eichl.)

COMMELINALES

The order Commelinales is in the eight systems variously composed as to families. The family Commelinaceae is the one here chiefly considered. The plants of this family have sepals well differentiated from the petals. *Tradescantia* and *Commelina* are common native genera of our region. *Rhoeo* and *Zebrina* are among several tropical genera cultivated as ornamentals.

Opinions differ as to the importance of major characters in this order. Engler assigns chief importance to mealy endosperm. He therefore places the family Pontederiaceae, despite its lack of sepaloid calyx, in the Commelinales. On the basis of petaloid perianth, however, the Pontederiaceae are included in the Liliales by Endlicher, Eichler, Hutchinson, Melchior, and Cronquist. Bentham and Hooker and Bessey include both the Commelinaceae and Pontederiaceae in the Liliales.

Commelina is a Linnaean name commemorating two Dutch botanists named Commelyn, and the genus designated was chosen to bear this name because of the two conspicuous petals of its flower. A third and less conspicuous petal commemorates, according to Linnaeus, a third Commelyn "who died before accomplishing anything in botany."

COMMELINACEAE

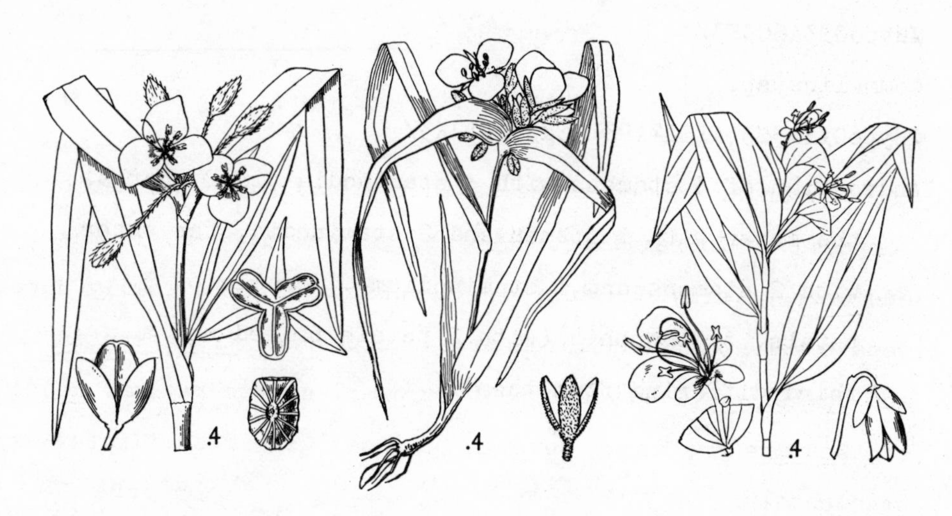

Tradescantia virginiana	Tradescantia bracteata	Commelina communis
-	-	-
spiderwort	spiderwort	dayflower

COMMELINACEAE

øHypoG2̸-3A6,2-3,SO,3-4,Ca,a̸,3K3/caps./oft succ.

Monocotyledonous plants with hypogynous flowers. Gynoecium of 2 to 3 united carpels. Androecium of 6, or 2 to 3, stamens; staminodia none, or 3 to 4. Corolla of 3 petals, actinomorphic or zygomorphic. Calyx of 3 sepals. Fruit a capsule. Plants often succulent.

Tradescantia sp.

a̸HypoGi̸3A6C3K3/

Commelina sp.

a̸HypoGi̸3A3S3(A1S2+A2S1)Ca̸3(2+1)K3/

Androecium of 3 stamens with 3 staminodia (in 2 whorls, the inner with 1 stamen and 2 staminodia, the outer with 2 stamens and 1 staminodium). Corolla of 3 petals, zygomorphic (with 2 petals paired and a third differing from them).

5 3:1

Endlicher places the Commelinales next to the Graminales, the first order in his monocot classification. Bentham and Hooker, however, include the Commelinaceae in their Liliales, the third order in their monocot sequence and one far from the Graminales. Eichler divides the monocot orders into two groups, and plants in the second of these usually have pistils with three functional carpels. Within this group the order Commelinales has first place. Engler adopts this group and retains the Commelinales in first position. Bessey follows Bentham and Hooker in including the Commelinaceae within the Liliales, the Liliales being the second order of the Monocotostrobiloididae, the first of his two monocot subclasses. Hutchinson, laying emphasis on the sepaloid calyx, makes the Commelinales the final order in the Calyciferidae, the first of his three monocot subclasses.

Melchior places the order Commelinales toward the middle of his monocot classification. It here stands among orders which follow the Liliales. Melchior states that the relationship between the Commelinales and Liliales is a remote one. Cronquist makes the Commelinales the first order in the Commelinidae, the second of his four monocot subclasses. He derives the Commelinidae from a line arising near the Alismatidae which after producing the offshoot Aridae continues as a stock which by branching gives rise to both the Commelinidae and Liliidae. Cronquist suggests that the ancestral form of these two groups would be referred to the Commelinidae. Cronquist's critical characters for the Commelinidae include: sepaloid sepals, starchy endosperm, and stomates with subsidiary cells. In his system the Commelinales become the pivotal order in monocot classification.

5 4:1

GLOSSARY FOR COMMELINALES

Commelinales. Order in class Monocotyledoneae.

Enantioblastae. Classis in sectio Amphibrya (Endl.)

Enantioblastae. Unterreihe in Abtheilung Monocotyledoneae (Eichl.)

Farinosae. Series in classis Monocotyledoneae (Engl.)
Commelinales. Order in division Calyciferae (Hutch.)
Commelinales. Reihe in Klasse Monocotyledoneae (Melch.)
Commelinales. Order in subclass Commelinidae (Cronq.)

Commelinaceae. Family in order Commelinales.
Commelynaceae. Ordo in classis Enantioblastae (Endl.)
Commelinaceae. Ordo in series Coronarieae (B.&H.)
Commelinaceae. Familie in Unterreihe Enantioblastae (Eichl.)
Commelinaceae. Familia in series Farinosae (Engl.)
Commelinaceae. Family in order Liliales (Bessey).
Commelinaceae. Family in order Commelinales (Hutch.), (Cronq.)
Commelinaceae. Familie in Reihe Commelinales (Melch.)

6 1:1

LILIALES

The order Liliales is commonly pivotal in the classification of monocots. The treatment of it often determines much of the character of monocot classification. In different systems the order varies in number of families. Only four families are here considered: the Juncaceae, Liliaceae, Amaryllidaceae, and Iridaceae. Only in the case of the type family is there agreement that it belongs in the order, and even here there are important differences as to the proper circumscription. The number of major genera in these four families is too large to permit enumeration here. Representative genera include: *Juncus*, *Asparagus*, *Lilium*, *Allium*, *Amaryllis*, and *Iris*.

Juncus is the ancient Latin name for the rush. *Lilium* is the ancient Latin name for the lily. *Amaryllis* is the name of a shepherdess in Virgil's *Eclogues*, from which Linnaeus took it for use as a plant name. Virgil had previously obtained it from poems of Theocritus. *Iris* is the ancient Greek name for the iris. The word has in Greek the earlier signification of "rainbow"; hence the literal meaning of the botanical name appears to be "rainbow flower," with reference to color transitions in the corolla.

6 2:1

JUNCACEAE

d̸HypoGi̸3L3,1,A6,3,P6(chaffy)/caps.

Monocotyledonous plants with hypogynous flowers. Gynoecium of 3 united carpels; locules 3, or 1. Androecium of 6 stamens, or 3. Perianth of 6 tepals (chaffy). Fruit a capsule.

Juncus sp.

d̸HypoGi̸3L1A6P6(chaffy)/

JUNCACEAE

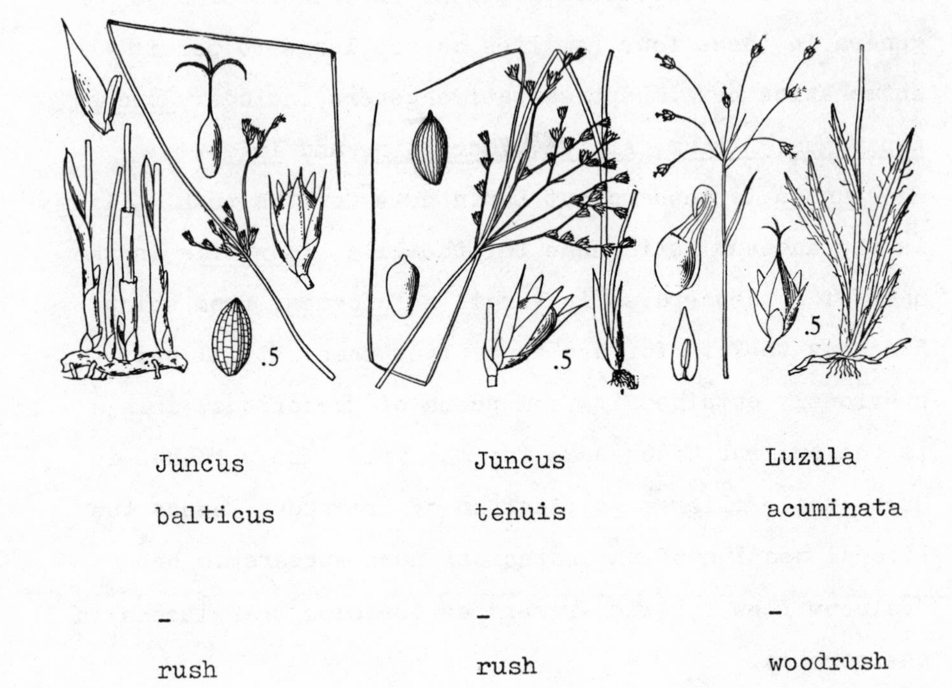

Juncus balticus	Juncus tenuis	Luzula acuminata
-	-	-
rush	rush	woodrush

LILIACEAE

d̸HypoG1̸3A(4-)6(Hypo,(epitep.),)Pi,1̸,(4-)6(corolloid)/

perf.,(dioec.)/alt.delim.(Cronq.)d̸Hypo,Epi

Monocotyledonous plants with hypogynous flowers. Gynoecium of 3 united carpels. Androecium of 6 stamens (or as few as 4) (hypogynous (or on some cases epitepalous)). Perianth of 6 tepals (or as few as 4), free or united (corolloid). Flowers perfect (or in some cases unisexual and the plants dioecious). By alternative delimitation

LILIACEAE

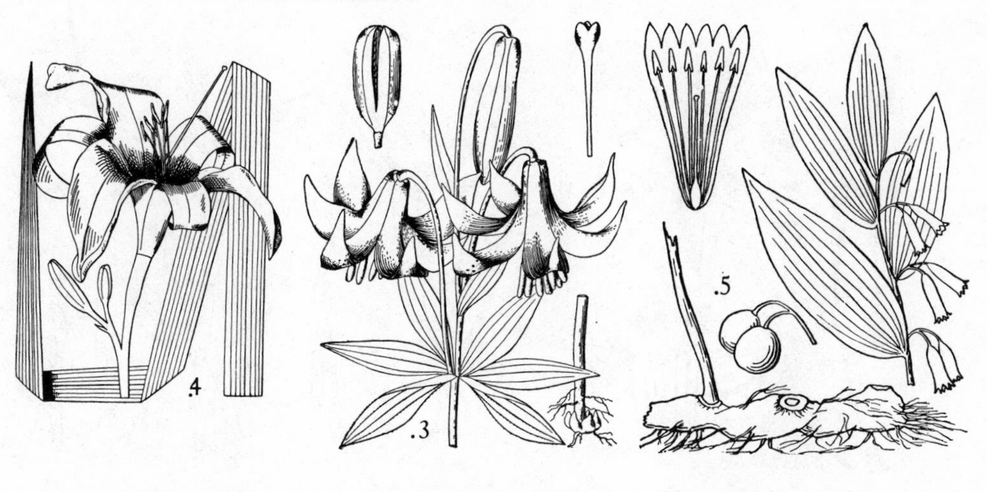

Hemerocallis fulva	Lilium canadense	Polygonatum canaliculatum
-	Canada	-
day lily	lily	Solomon's seal

(Cronquist): Monocotyledonous plants with hypogynous, or epigynous flowers.

Lilium sp.

d̸HypoGi̸3A6P6(corolloid)/

Polygonatum sp.

d̸HypoGi̸3A6(epitep.)Pi̸6(corolloid)/

AMARYLLIDACEAE

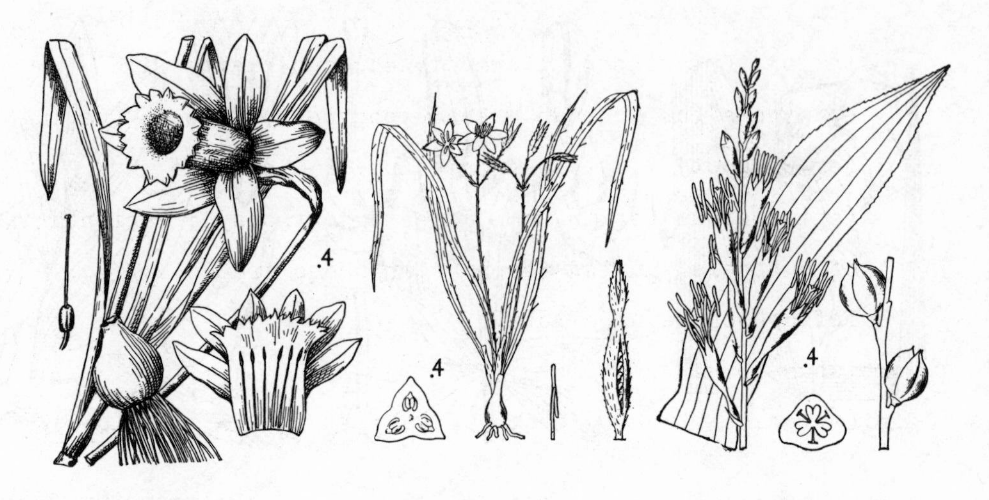

Narcissus pseudonarcissus	Hypoxis hirsuta	Agave virginica
daffodil	-	false
-	star grass	aloe

AMARYLLIDACEAE

d̸Epi(w.hypanth.tube oft.w.corona)Gi̸3A6(epitep.)Pi̸6 (corolloid)/caps.,ber./alt.delim.(Hutch.)d̸Epi,(Hypo) (w.,(not w.)...)...A6(epitep.,(Hypo))Pi̸,(i),6.../... infl.umbel.

Monocotyledonous plants with epigynous flowers (having a hypanthium tube often with a corona). Gynoecium of 3 united carpels. Androecium of 6 stamens (epitepalous). Perianth of 6 united tepals (corolloid). Fruit a capsule, or a berry. By alternative delimitation (Hutchinson): Monocotyledonous plants with epigynous (or in some cases hypogynous) flowers (having (or without) ...) ... Androecium of 6 stamens (epitepalous (or in some cases hypogynous)). Perianth of 6 tepals, united (or in some cases free) .../ ... Inflorescence umbellate.

Narcissus sp.

d̸Epi(w.hypanth.tube w.corona)Gi̸3A6(epitep.)P1̸6(corolloid)/

IRIDACEAE

d̸Epi(w.hypanth.tube)Gi̸3(styles oft.petaloid)A3Pa,a̸,6 (corolloid)/caps.

Monocotyledonous plants with epigynous flowers (having a hypanthium tube). Gynoecium of 3 united carpels (the styles often petaloid). Androecium of 3 stamens. Perianth of 6 tepals, actinomorphic or zygomorphic (corolloid). Fruit a capsule.

Iris sp.

d̸Epi(w.hypanth.tube)Gi̸3(styles petaloid)A3P6(corolloid)/

IRIDACEAE

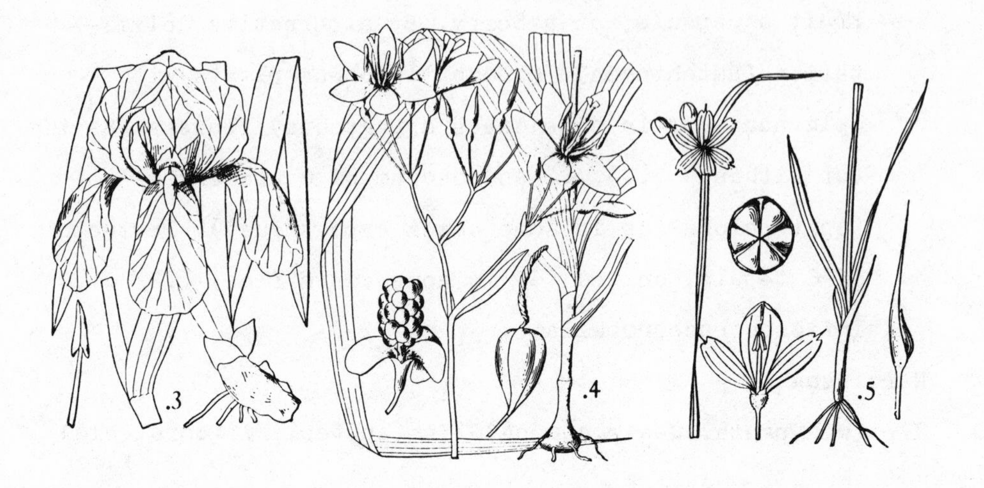

Iris germanica	Belamcanda chinensis	Sisyrinchium angustifolium
-	-	-
iris	blackberry lily	blue-eyed grass

6 3:1

Endlicher places the Liliales toward the middle of his sequence of monocot orders. Here they are directly before the Iridales and Orchidales. Bentham and Hooker place the Liliales after the Orchidales and Iridales, and this set of three orders commences their monocot classification. The order Liliales is placed third by them so that it can also be near to the order Juncales, which diverges from their Liliales by lacking a corolloid perianth.

Eichler divides his monocot orders into two groups, and pistils in the second of these usually have three functional carpels. Within this group the order Liliales stands between the Commelinales and Orchidales. Engler follows Eichler's classification in this regard. Bessey places the Liliales second only to the Alismatales in the Monocotostrobiloididae, the first of his two monocot subclasses. This indicates that he derives the Liliales from the Alismatales, a point on which his diagrams are explicit. All the other orders of monocots here considered he derives directly or indirectly from the Liliales. Hutchinson places the Liliales as the first order of the Corolliferidae, the second of his three monocot subclasses. He derives the Liliales together with all the other Corolliferidae orders from his first monocot subclass, the Calyciferidae. Melchior places the Liliales second only to the Alismatales in his sequence of monocot orders. His arrangement is harmonious with Besseyan patterns, but Melchior is reticent in affirming specific relationships among orders. Cronquist makes the Liliales the first order of the Liliidae, the fourth of his four monocot subclasses. He derives the Liliidae from a line arising near the Alismatidae, which after producing the offshoot Aridae continues as a stock which by branching gives rise to both the Commelinidae and Liliidae. Cronquist suggests that the ancestral form of these two groups would be referred to the Commelinidae. Cronquist's critical characters for the Liliidae include: petaloid calyx, nonstarchy endosperm, and stomates without subsidiary cells.

Endlicher accepts both the order Liliales, including the Juncaceae, and the order Iridales, including the Amaryllidaceae. Eichler, Engler, and Bessey accept the Liliales including the Juncaceae, but Bentham and Hooker, Hutchinson, Melchior, and Cronquist hold for a separate order Juncales. Cronquist alone, however, places his Juncales remote from the Liliales. As for Endlicher's Iridales, Bentham and Hooker and Bessey accept it. Hutchinson accepts it but splits off the Amaryllidaceae to form the order Amaryllidales. On the other hand Eichler, Engler, and Melchior reject the order and include both the Iridaceae and Amaryllidaceae in the Liliales. Cronquist goes one step further and lumps *Amaryllis* into the Liliaceae, but he retains the Iridaceae. Bessey places the Iridales in a different subclass from his Liliales. Hutchinson attacks the character of epigyny on which the family Amaryllidaceae has in part been based, and replaces it with an umbellate inflorescence character basic to his order Amaryllidales. As a consequence he transfers *Allium* from the Liliaceae and Liliales to the Amaryllidaceae and Amaryllidales. Cronquist concurs with Hutchinson's views on the inadequacy of the epigyny character, but he refuses to accept the umbellate inflorescence character as of ordinal or even familial value.

Within the Liliales Endlicher places the Juncaceae ahead of the Liliaceae, and within the Iridales he places the Iridaceae ahead of the Amaryllidaceae. Within the Iridales Bentham and Hooker follow Endlicher in placing the Iridaceae ahead of the Amaryllidaceae. Their Liliaceae and Juncaceae are in separate orders. Within the Liliales Eichler's sequence of families runs: Juncaceae, Liliaceae, Amaryllidaceae, Iridaceae. Engler adopts this Eichlerian sequence. Within the Liliales in his subclass Monocotostrobiloididae Bessey places the Liliaceae ahead of the Juncaceae, and within the Iridales in his subclass Monocotocotyloididae he places the Amaryllidaceae ahead of the Iridaceae. Hutchinson places the Liliaceae, Amaryllidaceae, and Iridaceae in separate orders within the subclass Corolliferidae. His Juncaceae are in still another subclass, the Glumifloridae. Within the Liliales Melchior's sequence runs: Liliaceae, Amaryllidaceae, and Iridaceae. His Juncaceae are in another order. Cronquist places the Juncaceae in an order of his subclass Commelinidae. His Liliaceae (incl. *Amaryllis*) precede the Iridaceae within the Liliales of his Liliidae.

GLOSSARY FOR LILIALES

Juncales. Order in class Monocotyledoneae.

Calycinae. Series in subdivisio Monocotyledones (B.&H.)

Juncales. Order in division Glumiflorae (Hutch.)

Juncales. Reihe in Klasse Monocotyledoneae (Melch.)

Juncales. Order in subclass Commelinidae (Cronq.)

Liliales. Order in class Monocotyledoneae.

Coronariae. Classis in sectio Amphibrya (Endl.)

Coronarieae. Series in subdivisio Monocotyledones (B.&H.)

Liliiflorae. Unterreihe in Abtheilung Monocotyledoneae (Eichl.)

Liliiflorae. Series in classis Monocotyledoneae (Engl.)

Liliales. Order in subclass Strobiloideae (Bessey).

Liliales. Order in division Corolliferae (Hutch.)

Liliiflorae. Reihe in Klasse Monocotyledoneae (Melch.)

Liliales. Order in subclass Liliidae (Cronq.)

Amaryllidales. Order in class Monocotyledoneae.

Amerillidales. Order in division Corolliferae (Hutch.)

Iridales. Order in class Monocotyledoneae.

Ensatae. Classis in sectio Amphibrya (Endl.)

Epigynae. Series in subdivisio Monocotyledones (B.&H.)

Iridales. Order in subclass Cotyloideae (Bessey).

Iridales. Order in division Corolliferae (Hutch.)

Juncaceae. Family in order Liliales.

Juncaceae. Ordo in classis Coronariae (Endl.)

Juncaceae. Ordo in series Calycinae (B.&H.)

Juncaceae. Familie in Unterreihe Liliiflorae (Eichl.)

Juncaceae. Familia in series Liliiflorae (Engl.)

Juncaceae. Family in order Liliales (Bessey).

Juncaceae. Family in order Juncales (Hutch.), (Cronq.)

Juncaceae. Familie in Reihe Juncales (Melch.)

Liliaceae. Family in order Liliales.

Liliaceae. Ordo in classis Coronariae (Endl.)

Liliaceae. Ordo in series Coronarieae (B.&H.)

Liliaceae. Familie in Unterreihe Liliiflorae (Eichl.)

Liliaceae. Familia in series Liliiflorae (Engl.)

Liliaceae. Family in order Liliales (Bessey), (Hutch.), (Cronq.)

Liliaceae. Familie in Reihe Liliiflorae (Melch.)

Amaryllidaceae. Family in order Liliales.

Amaryllideae. Ordo in classis Ensatae (Endl.)

Amaryllideae. Ordo in series Epigynae (B.&H.)

Amaryllidaceae. Familie in Unterreihe Liliiflorae (Eichl.)

Amaryllidaceae. Familia in series Liliiflorae (Engl.)

Amaryllidaceae. Family in order Iridales (Bessey).

Amaryllidaceae. Family in order Amaryllidales (Hutch.)

Amaryllidaceae. Familie in Reihe Liliiflorae (Melch.)

Iridaceae. Family in order Liliales.

Irideae. Ordo in classis Ensatae (Endl.)

Irideae. Ordo in series Epigynae (B.&H.)

Iridaceae. Familie in Unterreihe Liliiflorae (Eichl.)

Iridaceae. Familia in series Liliiflorae (Engl.)

Iridaceae. Family in order Iridales (Bessey), (Hutch.)

Iridaceae. Familie in Reihe Liliiflorae (Melch.)

Iridaceae. Family in order Liliales (Cronq.)

Commelinaceae. See 5 4.

Commelinaceae. Ordo in series Coronarieae (B.&H.)

Commelinaceae. Family in order Liliales (Bessey).

Marantaceae. See 7 4.

Scitamineae. Ordo in series Epigynae (B.&H.)

Marantaceae. Family in order Iridales (Bessey).

MARANTALES

The genus Thalia is the only representative of its order in our area. The order is not given regular coverage in the present survey, and to simplify nomenclatural complexities it is here termed Marantales.

Maranta is named in honor of the Venetian physician and botanist Bartolemmo Maranta. The name was first used in 1703 by Charles Plumier (Carolus Plumierus), a French missionary and botanist.

MARANTACEAE

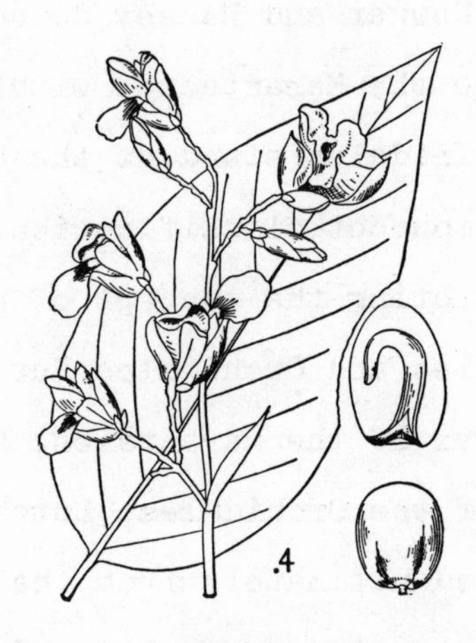

Thalia
dealbata

MARANTACEAE

d̸EpiGi̸3L3,1,A1S3-4(variously petaloid)Ci̸a̸3K3/bac.,caps.

Monocotyledonous plants with epigynous flowers. Gynoecium of 3 united carpels; the locules 3, or 1. Androecium of a single stamen, with staminodia 3 to 4 (variously petaloid). Corolla of 3 united petals, zygomorphic. Calyx of 3 sepals. Fruit baccate, or a capsule.

Thalia dealbata

d̸EpiGi̸3L1A1S3(variously petaloid)Ci̸a̸3K3/

Endlicher, Eichler, and Engler associate the Marantales with the Orchidales at the end of their monocot classifications. Bentham and Hooker and Bessey do not recognize the Marantales but place the Marantaceae within their Iridales. The Orchidales and Iridales stand at the beginning of the Bentham and Hooker monocot classification, but at the end of Bessey's. In rejecting the concept of a relationship between the Marantales and Orchidales Hutchinson lays stress on the sepaloid calyx of the Marantales, as opposed to the petaloid perianth of the Orchidales. Hutchinson diagrams the Marantales as derived ultimately from the Commelinales. Melchior retains the Englerian position for the Marantales and Orchidales at the end of his monocot classification, but he specifies that these orders have had parallel development rather than a direct relationship to each other. Cronquist follows Hutchinson in making a complete separation of the Marantales from the Orchidales. He makes the Marantales an offshoot from a line leading to the Commelinales.

GLOSSARY FOR MARANTALES

Marantales. Order in class Monocotyledoneae.

Scitamineae. Classis in sectio Amphibrya (Endl.)

Scitamineae. Unterreihe in Abtheilung Monocotyledoneae (Eichl.)

Scitamineae. Series in classis Monocotyledoneae (Engl.)

Zingiberales. Order in division Calyciferae (Hutch.)

Scitamineae. Reihe in Klasse Monocotyledoneae (Melch.)

Zingiberales. Order in subclass Commelinidae (Cronq.)

Marantaceae. Family in order Marantales.

Cannaceae. Ordo in classis Scitamineae (Endl.)

Scitamineae. Ordo in series Epigynae (B.&H.)

Marantaceae. Familie in Unterreihe Scitamineae (Eichl.)

Marantaceae. Familia in series Scitamineae (Engl.)

Marantaceae. Family in order Iridales (Bessey).

Marantaceae. Family in order Zingiberales (Hutch.), (Cronq.)

Marantaceae. Familie in Reihe Scitamineae (Melch.)

ORCHIDALES

The Burmanniaceae often and the Hydrocharitaceae seldom have been placed in the Orchidales, but only the Orchidaceae are here considered. Genera of this family represented in our area include Cypripedium, Orchis, and Calopogon. The number of cultivated genera is large and includes Cattleya, Epidendrum, and Vanilla.

Orchis is the ancient Greek name for the orchid and signifies "testicle" in allusion to the tubers of this genus. The Greek word is not a d-stem, but it was so treated by Linnaeus; hence the d in Orchidaceae and Orchidales. On the basis of these forms, Lindley in 1845 coined the English word orchid. Fernald, however, does not recognize orchid as the common name, but cites orchis as maintaining its place in vernacular English.

ORCHIDACEAE

ØEpiG/3L1A1-2(gynandrous)S2-0PØ6(corolloid)/caps.

Monocotyledonous plants with epigynous flowers. Gynoecium of 3 united carpels and a single locule. Androecium of 1 to 2 stamens (gynandrous); staminodia 2 to none. Perianth of 6 tepals, zygomorphic (corolloid). Fruit a capsule.

Orchis sp.

ØEpiG/3L1A1(gynandrous)PØ6(corolloid)/

Cypripedium sp.

ØEpiG/3L1A2(gynandrous)S1PØ6(2+1)+(1+/2)/

ORCHIDACEAE

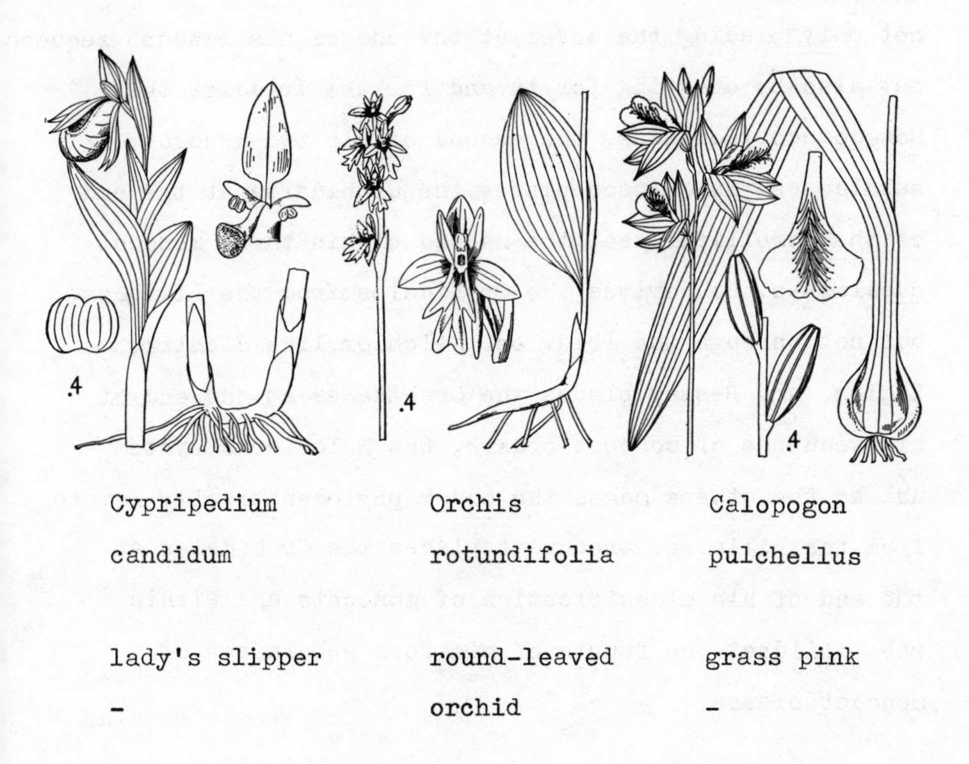

Cypripedium candidum	Orchis rotundifolia	Calopogon pulchellus
lady's slipper	round-leaved orchid	grass pink
-		-

8 3:1

Endlicher places the Orchidales after the Liliales and Iridales at a more or less central position in his sequence of monocot orders. Bentham and Hooker place the Orchidales at the beginning of their monocot sequence. This position suggests that they regard the Orchidales as isolated from monocot orders other than the Iridales and Liliales which follow in the sequence. Eichler places the Orchidales after the Liliales (incl. Iridaceae) and at the end of his monocot sequence. Engler's system follows Eichler in this. Bessey gives emphasis to orchidalian advancement by not only placing the order at the end of his monocot sequenc but also by creating for it and for the Iridales the Monocotocotyloididae, the second of his two monocot subclasses. Hutchinson places the Orchidales at the end of the Corolliferidae, the second of his three monocot subclasses. He derives the Orchidales from the Liliales, but not through the Iridales. Melchior like Eichler, Engler, and Bessey places the Orchidales at the end of his sequence of monocot orders, but Melchior's system unlike the others makes the order phylogenetically remote from the Liliales. Cronquist places the Orchidales at the end of his classification of monocots and within the Liliidae, the fourth of his four subclasses of monocot orders.

8 4:1

GLOSSARY FOR ORCHIDALES

Orchidales. Order in class Monocotyledoneae.

Gynandrae. Classis in sectio Amphibrya (Endl.)

Microspermae. Series in subdivisio Monocotyledones (B.&H.)

Gynandrae. Unterreihe in Abtheilung Monocotyledoneae (Eichl.)

Microspermae. Series in classis Monocotyledoneae (Engl.)

Orchidales. Order in subclass Cotyloideae (Bessey).

Orchidales. Order in division Corolliferae (Hutch.)

Microspermae. Reihe in Klasse Monocotyledoneae (Melch.)

Orchidales. Order in subclass Liliidae (Cronq.)

Orchidaceae. Family in order Orchidales.

Orchideae. Ordo in classis Gynandrae (Endl.)

Orchideae. Ordo in series Microspermae (B.&H.)

Orchidaceae. Familie in Unterreihe Gynandrae (Eichl.)

Orchidaceae. Familia in series Microspermae (Engl.)

Orchidaceae. Family in order Orchidales (Bessey), (Hutch.), (Cronq.)

Orchidaceae. Familie in Reihe Microspermae (Melch.)

DICOTYLEDON SUBCLASSES

Endlicher divides the class Dicotyledoneae into the subclasses: Apetalidae, Gamopetalidae, and Polypetalidae. Bentham and Hooker recognize the same subclasses, but they reverse the sequence to: Polypetalidae, Gamopetalidae, Apetalidae. Eichler unites the Apetalidae and Polypetalidae into the Agamopetalidae which follows his Gamopetalidae. In adopting the Agamopetalidae, Eichler follows the 1843 system of the paleobotanist Adolphe Brongniart who rejected the Apetalidae on the grounds that the taxa of this subclass were in fact derived from petaliferous forms. Brongniart advocated assigning apetalous taxa to putatively related petaliferous taxa. As an abstract idea this proposal has been widely accepted, but the puting it into practice has resulted in much phylogenetic speculation and disagreement. Eichler, however, rejects the Brongniart proposal itself and argues that the apetalous condition is commonly primitive. Within his Agamopetalidae he separates apetalous from polypetalous taxa. Eichler later reversed his sequence of subclasses, so they run in his 1883 system: Agamopetalidae, Gamopetalidae. Engler and Melchior adopt the latter sequence. The systems of Endlicher, Bentham and Hooker, Eichler, Engler, and Melchior are comparable as to their dicot subclasses. Those of Bessey, Hutchinson, and Cronquist, however, are fundamentally different not only from the subclasses of the five systems enumerated above but also from those of each other. Bessey's subclasses may here be termed the Dicotostrobiloididae and the Dicotocotyloididae. The former is interpreted as fundamentally hypogynous, and the latter is interpreted as fundamentally perigynous to epigynous. Hutchinson's subclasses emphasize plant tissues, classifying orders as fundamentally woody or as fundamentally herbaceous. The subclasses are here termed the Lignosidae and the Herbacidae. Cronquist's dicot subclasses are six in number. These are not given descriptive names such as Gamopetalidae, Dicotostrobiloididae, or Lignosidae, but names derivative from orders. These subclasses are here termed: Magnoliidae, Hamamelidae, Caryophyllidae, Theidae, Rosidae, and Asteridae. These subclasses cannot be sharply defined. Technical characters are employed to make them correspond to putatively phylogenetic entities. Definition by enumeration will be provided in connection with the discussion of specific subclasses and orders.

GLOSSARY FOR DICOTYLEDON SUBCLASSES, ETC.

Apetalae, Gamopetalae, Dialypetalae. Cohortes in sectio Acramphibrya (Endl.)

Polypetalae, Gamopetalae, Monochlamydeae. Subclasses in classis Angiospermae (B.&H.) In subdivisio Dicotyledones.

Chori- und Apetalae, Sympetalae. Klassen in Abtheilung Dicotyledoneae (Eichl.)

Archichlamydeae, Metachlamydeae. Subclasses in classis Dicotyledoneae (Engl.)

Strobiloideae, Cotyloideae. Subclasses in class Oppositifoliae (Bessey).

Lignosae, Herbaceae. Divisions in subphylum Dicotyledones (Hutch.)

Archichlamydeae, Sympetalae. Unterklassen in Klasse Dicotyledoneae (Melch.)

Magnoliidae, Hamamelidae, Caryophyllidae, Dilleniidae, Rosidae, Asteridae. Subclasses in class Magnoliatae (Cronq.)

APETALOUS ORDER GROUPS

Endlicher's orders of the subclass Apetalidae have the sequence: (1) Urticales (incl. Fagaceae, Platanaceae, Salicaceae), (2) Chenopodiales (incl. Polygonaceae), (3) Elaeagnales. The first order has unisexual flowers as a character and is predominantly woody, while the second has mostly bisexual flowers and is predominantly herbaceous. Endlicher recognizes the order Caryophyllales within his subclass Polypetalidae. Within the order Rutales of this subclass he places the family Juglandaceae.

Bentham and Hooker's orders of the subclass Apetalidae have the sequence: (1) Chenopodiales (incl. Polygonaceae), (2) Elaeagnales, (3) Urticales (incl. Euphorbiaceae, Platanaceae, Juglandaceae, Fagaceae), (4) Salicales. These correspond to Endlicher's orders except that the Salicales are here split off from the Urticales. Bentham and Hooker recognize the order Caryophyllales within their subclass Polypetalidae.

Eichler's apetalous order group of the subclass Agamopetalidae has the sequence: (1) Fagales (incl. Juglandaceae, Salicaceae), (2) Urticales (incl. Platanaceae), (3) Caryophyllales (incl. Polygonaceae). In thus splitting off the Fagales from the Urticales, Eichler recognizes the amentiferous group at ordinal rank.

Engler's apetalous order group of the subclass Agamopetalidae has the sequence: (1) Salicales, (2) Juglandales, (3) Fagales, (4) Urticales, (5) Polygonales, (6) Caryophyllales. Engler here distributes the taxa of three Eichlerian orders into six orders.

Bessey, following Brongniartian theory, assigns apetalous taxa to putatively related petaliferous ones. Part of the apetalous taxa fall within his subclass Dicotostrobiliodidae. Here he places the Urticaceae within the Malvales. In the Caryophyllales he includes not only the Polygonaceae, but also the Salicaceae—disturbing the erstwhile character of the Caryophyllales as a predominantly herbaceous group. Part of the apetalous taxa fall within his subclass Dicotocotyloididae. Here he places the Juglandaceae and Fagaceae within his Sapindales.

In Hutchinson's system part of the apetalous taxa fall within his subclass Lignosidae. These are in two series in his diagrams. One is amentiferous and includes the Salicales, Fagales, and Juglandales. The other consists of the Urticales. Both series are derived from the Hamamelidales. Part of the apetalous taxa fall within his subclass Herbacidae. These are in a single series in his diagrams. Here the Caryophyllales, taken as a fully petaliferous taxon, give rise to the Polygonales; and they in turn give rise to the Chenopodiales.

Melchior's apetalous order group of the subclass Agamopetalidae follows that of Engler with modifications. At the beginning of the order group Melchior places the Juglandales ahead of the Salicales. He notes a diversity of views as to the origin of the Salicales and that a consensus holds them to be advanced. He finds, however, that such a consensus does not help much in determining a phylogenetic position for the order. Similar considerations apply to his other apetalous orders.

In Cronquist's system most of the apetalous taxa fall within the Hamamelidae, the second of his six dicot subclasses. These include the Urticales, Juglandales, and Fagales—which are diagrammed as separately derived from the Hamamelidales (incl. Platanaceae). The Caryophyllales are included within the Caryophyllidae, the third of his six dicot subclasses. Here the subclass and order are taken as fundamentally petaliferous, but they include several apetalous families. The Salicales are included in the Theidae, the fourth of his six dicot subclasses. Cronquist presents as a phylogenetic sequence: Theales, Violales, Salicales. In placing the Salicales after the Violales he emphasizes parietal placentation and capsular structure of the ovary.

GLOSSARY FOR APETALOUS SUPERORDERS, ETC.

Juliflorae (Unterreihen: Piperinae, Amentaceae, Urticinae). Reihe in Klasse Chori- und Apetalae (Eichl.)

Centrospermae (Unterreihe: Centrospermae (sic)). Reihe in Klasse Chori- und Apetalae (Eichl.)

Hysterophyta (Familien: Aristolochiaceae, Santalaceae). Anhang in Klasse Chori- und Apetalae (Eichl.) Analogously with Centrospermae (above) the Anhang Hysterophyta functions at both the Reihe and Unterreihe levels.

SAURURALES

The genus Saururus, lizardtail, is the only representative of its order in our area. The order is not given regular coverage in the present survey, and to simplify nomenclatural complexities it is here termed Saururales.

Saururus has extreme floral simplicity, the bisexual flowers lacking perianth. There are 3 to 4 carpels united at the base and 6 to 8 stamens. The flowers are in dense spikes.

The name Saururus is a modern compound noun composed of the Greek roots for "lizard" and "tail," and alludes to the spikes.

SAURURACEAE

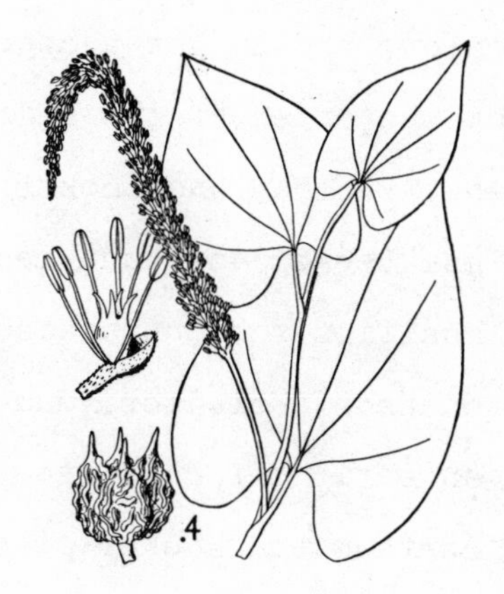

Saururus

cernuus

-

lizard's tail

SAURURACEAE

dHypoGi,i̸,3-4A(2-)6-8/indeh.spike

Dicotyledonous plants with hypogynous flowers. Gynoecium
of 3 to 4 distinct, or united carpels. Androecium

of 6 to 8 stamens (or as few as 2). Fruit indehiscent. Inflorescence a spike.

Saururus cernuus

dHypoG_4A6/

Gynoecium of 4 carpels united at the base.

9 3:1

Endlicher, Eichler, and Engler accept the Saururales as primitive and place the order at the beginning of their dicot classification. Bentham and Hooker treat it as having uncertain affinities. Bessey does not recognize the Saururales, but classifies the Saururaceae within his order Ranales. Hutchinson, Melchior, and Cronquist treat the Saururales as an early offshoot from ranalian stock.

Endlicher and Bentham and Hooker classify the Saururales in the subclass Apetalidae. Eichler and Engler classify the order in the apetalous portion of the subclass Agamopetalidae. Melchior, however, associates it with the polypetalous portion of the subclass Agamopetalidae. Hutchinson classifies the order in the subclass Herbacidae. Cronquist classifies the order in the subclass Magnoliidae.

GLOSSARY FOR SAURURALES

Saururales. Order in subclass Agamopetalidae.

Piperitae. Classis in cohors Apetalae (Endl.)

Micrembryeae. Series in subclassis Monochlamydeae (B.&H.)

Piperinae. Unterreihe in Reihe Juliflorae (Eichl.)
In Klasse Chori- und Apetalae.

Piperales. Series in subclassis Archichlamydeae (Engl.)

Piperales. Order in division Herbaceae (Hutch.)

Piperales. Reihe in Unterklasse Archichlamydeae (Melch.)

Piperales. Order in subclass Magnoliidae (Cronq.)

Saururaceae. Family in order Saururales.

Saurureae. Ordo in classis Piperitae (Endl.)

Piperaceae. Ordo in series Micrembryeae (B.&H.)

Saururaceae. Familie in Unterreihe Piperinae (Eichl.)

Saururaceae. Familia in series Piperales (Engl.)

Saururaceae. Family in order Ranales (Bessey).

Saururaceae. Family in order Piperales (Hutch.), (Cronq.)

Saururaceae. Familie in Reihe Piperales (Melch.)

SALICALES

The order Salicales consists of the single family Salicaceae, which contains only the genera <u>Salix</u> and <u>Populus</u>. These genera include several species prominent in the natural vegetation of our area and also a number of introduced ornamentals appropriate where rapid growth is desired. <u>Salix alba</u>, white willow, and <u>Salix babylonica</u>, weeping willow, are introduced forms of large size. <u>Populus deltoides</u>, cottonwood, is a native tree with spreading habit. <u>Populus nigra</u> var. <u>italica</u>, Lombardy poplar, is a cultivated staminate form. Its branches ascend nearly parallel to the trunk to form a distinctive columnar tree top.

<u>Salix</u> is the ancient Latin name for the willow. The word is cognate with the English noun <u>sallow</u>, an old name for "willow." This relationship implies that the word <u>Salix</u> has an Indo-European origin anterior to classical antiquity.

10 2:1

SALICACEAE

dPsiloG/2-4L1A2-n/glan.1,2,q/Bl/dioec.caps.

Dicotyledonous plants with psilogynous flowers. The

SALICACEAE

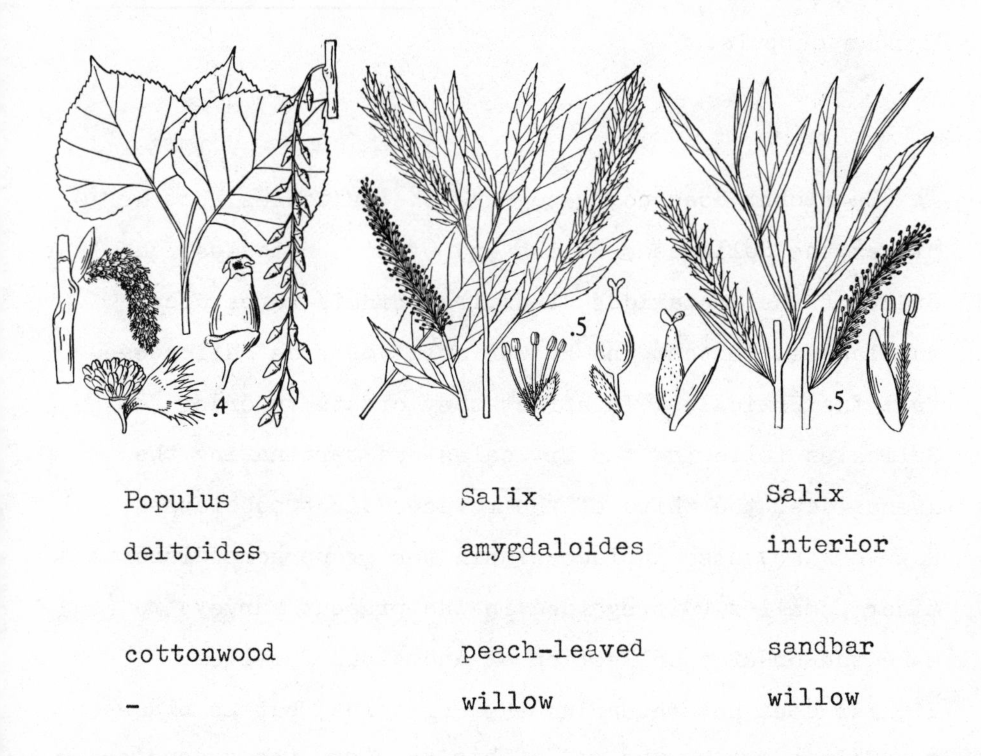

Populus deltoides	Salix amygdaloides	Salix interior
cottonwood	peach-leaved willow	sandbar willow
-		

gynoecium of 2 to 4 united carpels and a single locule. Androecium of 2 to many stamens. Glands 1, or 2, or the morphological number not evident. Bract 1. Plants dioecious. The fruit a capsule.

Salix sp.

dPsiloG̸2L1A5/glan.1(f),2(m)/dioec.

Glands 1 (in pistillate flowers), or 2 (in staminate flowers).

Populus sp.

dPsiloG/2L1A6-12/glan.cuppule/B1/dioec.

Gland a cuppule.

10 3:1

Endlicher does not recognize the Salicales, but he places the Salicaceae at the end of his Urticales, the last order of the Apetalidae, the first of his three dicot subclasses. Bentham and Hooker split off the Salicaceae from the Urticales. In effect they create an order Salicales following the Urticales and terminating the Apetalidae, the third of their three dicot subclasses. However they also include within the group some discordant minor families (disregarded in the present survey) to make the order a collection of anomalous families. Eichler does not recognize the Salicales, but he places the Salicaceae at the end of his Fagales, the second order of his apetalous group within the Agamopetalidae, the second of his two dicot subclasses. Eichler's order Fagales is characterized by amentiferous inflorescences, and it has often been condemned as failing to recognize that these families may be advanced by reduction. Eichler, however, recognizes the glands of the Salix flower and the cupule of the Populus flower as rudimentary perianths. Engler fragments Eichler's amentiferous Fagales and recognizes the Salicales. But whereas the Salicaceae come as last of the families within Eichler's order, the Salicales stand first among the new orders which Engler creates from that group. Thus the order follows directly after the Saururales in the Agamopetalidae, the first of Engler's two dicot subclasses. Bessey does not recognize the Salicales. He alone includes the Salicaceae in the Caryophyllales, the last order of the Dicotostrobiloididae, the first of his two dicot subclasses. The single locule of the Salicaceae is of importance in this classification. It must be observed that within the Caryophyllales Bessey's family Salicaceae follows the Tamaricaceae, both families having seeds with tufts of hair which allow dissemination by wind transport. Now the family Tamaricaceae is placed in association with the Thealian complex by Endlicher, Eichler, Hutchinson, Melchior, and Cronquist; and some of them derive the Salicaceae from this complex. Bentham and Hooker, however, whom Bessey has followed in assigning the Tamaricaceae to the Caryophyllales themselves explicitly reject the concept of affinity between the Tamaricaceae and Salicaceae. Hutchinson, far from following Bessey by including the Salicaceae in the Caryophyllales, assigns the orders Salicales and Caryophyllales to separate subclasses. The Salicales he puts in the Lignosidae, the first of his two dicot subclasses. Within this subclass he associates the Salicales with the Fagales, Juglandales, and other orders to form an aggregation analogous to the amentiferous groups of some other systems. He proposes as a phylogenetic sequence: Hamamelidales, Salicales. Melchior finds the theory of parietalic thealian origin for the Salicaceae more attractive than alternative hamamelidalian, caryophyllalian, and papaveralian theories. He refrains, however, from adopting any such theory and leaves the Salicales associated in an amentiferous order group within the Agamopetalidae, the first of his two dicot subclasses. This position is based on general morphological similarities devoid of phylogenetic theory. Cronquist assigns the Salicales to the Theidae, the fourth of his six dicot subclasses. He diagrams as a phylogenetic sequence: Theales, Violales, Salicales.

10 4:1

GLOSSARY FOR SALICALES

Salicales. Order in subclass Agamopetalidae.

Ordines anomali. Series in subclassis Monochlamydeae (B.&H.)

Salicales. Series in subclassis Archichlamydeae (Engl.)

Salicales. Order in division Lignosae (Hutch.)

Salicales. Reihe in Unterklasse Archichlamydeae (Melch.)

Salicales. Order in subclass Dilleniidae (Cronq.)

Salicaceae. Family in order Salicales.

Salicineae. Ordo in classis Juliflorae (Endl.)

Salicineae. Ordo in series Ordines anomali (B.&H.)

Salicineae. Familie in Unterreihe Amentaceae (Eichl.)

Salicaceae. Familia in series Salicales (Engl.)

Salicaceae. Family in order Caryophyllales (Bessey).

Salicaceae. Family in order Salicales (Hutch.), (Cronq.)

Salicaceae. Familie in Reihe Salicales (Melch.)

MYRICALES

Myrica construed broadly is the only genus of its order in our area, and because of this small representation the order is not given regular coverage in the present survey. The plants are shrubs or trees with simple or pinnately incised leaves having aromatic glands. Some of the species which occur naturally in our area are also cultivated as ornamental shrubs. These include *M. gale*, sweet gale, and *M. pensylvanica*, bayberry.

The name *Myrica* is the Latinized form of ancient Greek *myrikē*, a name for tamarisk. Since however the ancient Latin name *tamarix* had already been used to provide the generic name for tamarisk, *myrikē* was appropriated for use in a new signification.

MYRICACEAE

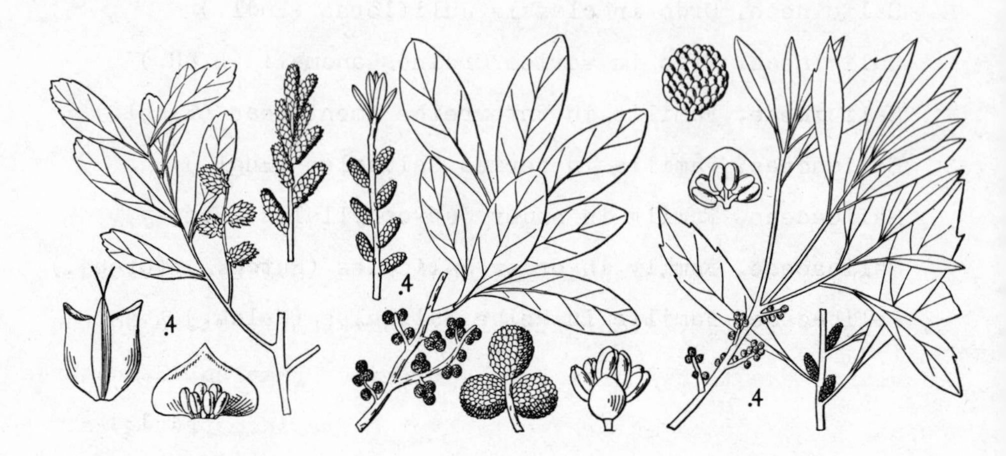

Myrica gale	Myrica pensylvanica	Myrica cerifera
sweet gale	bayberry	wax myrtle
-	-	-

MYRICACEAE

dPsilo,Hypo,G/2L1A(2-)4-8(-n)B2-8/monoec.,dioec.,bisex., drup.ament.

Dicotyledonous plants with psilogynous, or hypogynous, flowers. Gynoecium of 2 united carpels and a single locule. Androecium of 4 to 8 stamens (or as few as 2, or as many as more than 10). Bracts 2 to 8. Plants monoecious, or dioecious, or flowers bisexual. Fruit drupaceous. Inflorescences aments.

Myrica cerifera

dPsiloG*/*2L1A4B4/monoec.

11 3:1

Endlicher and Bentham and Hooker include the Myricaceae in the Urticales. Eichler includes the family close to the Juglandaceae in his broadly amentiferous Fagales. Structures usually accounted as bracts in Myrica are interpreted as sepals by Eichler. Engler provides for the Myricaceae the coextensive order Myricales. He associates the order with other orders which have psilogynous or hypogynous flowers. Bessey includes the Myricaceae in his Sapindales, where they stand after the Juglandaceae, Betulaceae, and Fagaceae. Hutchinson includes the Myricales in the Lignosidae, the first of his two dicot subclasses, and he associates the order with amentiferous orders of putative hamamelidalian derivation. Melchior includes the Myricaceae in his Juglandales. Cronquist places the Myricales in his amentiferous Hamamelidae, the second of his six dicot subclasses. He diagrams as a phylogenetic sequence: Hamamelidales, Myricales.

GLOSSARY FOR MYRICALES

Myricales. Order in subclass Agamopetalidae.

Myricales. Series in subclassis Archichlamydeae (Engl.)

Myricales. Order in division Lignosae (Hutch.)

Myricales. Order in subclass Hamamelidae (Cronq.)

Myricaceae. Family in order Myricales.

Myriceae. Ordo in classis Juliflorae (Endl.)

Myricaceae. Ordo in series Unisexuales (B.&H.)

Myricaceae. Familie in Unterreihe Amentaceae (Eichl.)

Myricaceae. Familia in series Myricales (Engl.)

Myricaceae. Family in order Sapindales (Bessey).

Myricaceae. Family in order Myricales (Hutch.), (Cronq.)

Myricaceae. Familie in Reihe Juglandales (Melch.)

12 1:1

LEITNERIALES

The species Leitneria floridana is the only representative of its order, and in our area occurs only in Missouri. Because of this small representation, the order is not given regular coverage in the present survey. The plants are shrubs or trees with simple entire leaves. Their habitat is in swampy areas of the Southern United States. The wood is of exceptionally light weight.

The name Leitneria commemorates E. F. Leitner, a physician and naturalist who died in Florida in 1838. A. W. Chapman published this name in his Flora of the Southern United States (1860). It is thus too late for Endlicher (died 1849) to classify.

12 2:1

LEITNERIACEAE

Leitneria

floridana

-

corkwood

LEITNERIACEAE

dHypoGlA1-4(in cyme3-12)P3-4(f)(scales)O(m)/dioec.drup.
ament./vs.dPsiloPO(f)B3-4(f)(scales)

Dicotyledonous plants with hypogynous flowers. Gynoecium of a single carpel. Androecium of 1 to 4 stamens (appearing in cymes having 3 to 12 stamens). Perianth of 3 to 4 tepals (in pistillate flowers) (scales) or none (in staminate flowers). Plants dioecious. Fruit drupaceous. Inflorescences aments. By alternative interpretation: Dicotyledonous plants with psilogynous flowers. Perianth none (in pistillate flowers). Bracts 3 to 4 (in pistillate flowers) (scales).

Leitneria floridana

dHypoGlA4(in cyme 12)P4(f)(scales)O(m)/dioec.

12 3:1

Leitneria is a genus unknown to Endlicher. Bentham and Hooker place the Leitneriaceae in the Urticales. Eichler does not recognize the Leitneriaceae, but places Leitneria in the family Myricaceae. Engler accords the genus both familial and ordinal status. His Leitneriales follow the Myricales. Bessey severs the myricalian association and places the Leitneriaceae in the Magnoliales. Hutchinson diagrams the Leitneriales as derived from the Hamamelidales, but he himself casts doubt on this idea and suggests an alternative derivation from the Rosales. Melchior mentions theories based on vascular anatomy suggesting a rosalian or geranialian origin for the Leitneriales. He himself, however, retains the Leitneriales within an amentiferous order group, a conservative morphological position in lieu of a phylogenetic theory. Cronquist follows Hutchinson in diagramming the Leitneriales as derived from the Hamamelidales; but he does so with diffidence, noting that "this possibility seems as likely as any." There is a modern consensus that the Leitneriales are not only advanced, but so far advanced that their phylogeny is obscure.

12 4:1

GLOSSARY FOR LEITNERIALES

Leitneriales. Order in subclass Agamopetalidae.
  Leitneriales. Series in subclassis Archichlamydeae (Engl.)
  Leitneriales. Order in division Lignosae (Hutch.)
  Leitneriales. Reihe in Unterklasse Archichlamydeae (Melch.)
  Leitneriales. Order in subclass Hamamelidae (Cronq.)
Leitneriaceae. Family in order Leitneriales.
  Leitnerieae. Ordo in series Unisexuales (B.&H.)
  Leitneriaceae. Familia in series Leitneriales (Engl.)
  Leitneriaceae. Family in order Ranales (Bessey).

Leitneriaceae. Family in order Leitneriales (Hutch.), (Cronq.)

Leitneriaceae. Familie in Reihe Leitneriales (Melch.)

JUGLANDALES

The order Juglandales consists of the single family Juglandaceae, which in our area is represented only by the genera *Juglans* and *Carya*. These genera include several species which are prominent in our natural vegetation and which are also commonly cultivated as ornamentals and for their wood and nuts. *Juglans nigra*, black walnut, and *Juglans cinerea*, butternut, are native plants. *Juglans regia*, English walnut, is cultivated in our area. *Carya illinoensis*, pecan, is the largest of the hickories and is important for its nuts. *Carya ovata*, shag-bark hickory, is valued for wood used in tool handles and for its hickory nuts.

Juglans is the ancient Latin name for the walnut. It is derived from the components *Ju-*, "Jupiter," and *glans* "acorn, chestnut, walnut." It is a loan translation of the Greek *Diosbalanos*, a word having two components cognate with those of *Juglans*.

JUGLANDACEAE

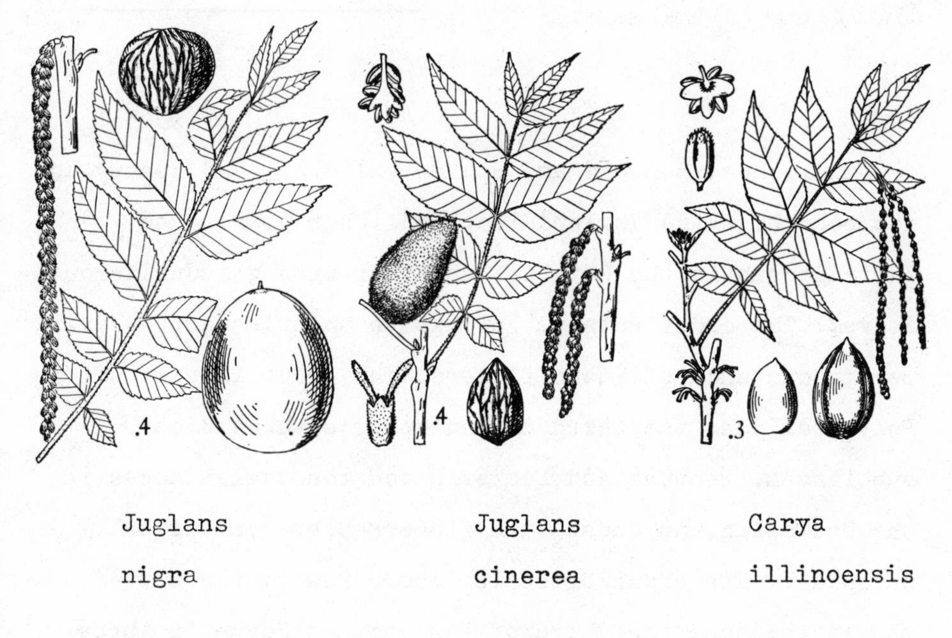

Juglans	Juglans	Carya
nigra	cinerea	illinoensis
black	butternut	pecan
walnut	-	-

JUGLANDACEAE

dEpiGi̸2-4A3-nPi̸3-6/monoec.drup.(hypanth.+exocarp husk)
ament.(m)

Dicotyledonous plants with epigynous flowers. Gynoecium of 2 to 4 united carpels. Androecium of 3 to many stamens. Perianth of 3 to 6 united tepals. Plants monoecious. Fruit drupaceous (the hypanthium and the exocarp forming a husk). Staminate inflorescences aments.

Juglans sp.

dEpiG/2AnP4(f)3(m)/monoec.

13 3:1

Endlicher, Bentham and Hooker, Eichler, and Bessey do not recognize the Juglandales. Endlicher places the Juglandaceae in the Rutales, an order with glanduliferous leaves. The order Rutales belongs to an order group having perigynous and epigynous flowers. The group is part of the Polypetalidae, the third of Endlicher's three dicot subclasses. Bentham and Hooker place the Juglandaceae in the Urticales, an order with flowers characteristically unisexual. The order Urticales comes toward the end of the Apetalidae, the third of Bentham and Hooker's three dicot subclasses. Eichler places the Juglandaceae in the Fagales, an order with inflorescences aments. The order Fagales comes toward the beginning of the Agamopetalidae, the second of Eichler's two dicot subclasses. Engler splits off the family Juglandaceae from the Fagales and raises it to ordinal status. The Juglandales differ from other amentiferous orders in having not only ovary inferior but also leaves compound. Engler retains the taxon toward the beginning of the Agamopetalidae, the first of his two dicot subclasses. Bessey places the Juglandaceae, along with other amentiferous taxa, in the Sapindales; and there they directly follow the Anacardiaceae. The order Sapindales comes toward the middle of the Dicotocotyloididae, the second of his two dicot subclasses. Hutchinson at one time derived the Juglandales from the Anacardiaceae of the Sapindales. He later rejected this idea, and includes the Juglandales in a hamamelidalian order group of the Lignosidae, the first of his two dicot subclasses. There it follows and is derived from the family Betulaceae of the order Fagales. Melchior places the Juglandales early in an amentiferous order group within the Agamopetalidae, the first of his two dicot subclasses. The order includes both the Myricaceae and Juglandaceae. Melchior recognizes as possible a derivation of the Juglandales from the Anacardiaceae, but he refrains from adopting the idea. Cronquist places the Juglandales in the Hamamelidae, the second of his six dicot subclasses. He diagrams as a phylogenetic sequence: Hamamelidales, Juglandales.

GLOSSARY FOR JUGLANDALES

Juglandales. Order in subclass Agamopetalidae.

Juglandales. Series in subclassis Archichlamydeae (Engl.)

Juglandales. Order in division Lignosae (Hutch.)

Juglandales. Reihe in Unterklasse Archichlamydeae (Melch.)

Juglandales. Order in subclass Hamamelidae (Cronq.)

Juglandaceae. Family in order Juglandales.

Juglandeae. Ordo in classis Terebinthineae (Endl.)

Juglandeae. Ordo in series Unisexuales (B.&H.)

Juglandaceae. Familie in Unterreihe Amentaceae (Eichl.)

Juglandaceae. Familia in series Juglandales (Engl.)

Juglandaceae. Family in order Sapindales (Bessey).

Juglandaceae. Family in order Juglandales (Hutch.), (Cronq.)

Juglandaceae. Familie in Reihe Juglandales (Melch.)

Myricaceae. See 11 4.

Myricaceae. Familie in Reihe Juglandales (Melch.)

14 1:1

FAGALES

The order Fagales consists of the families Betulaceae and Fagaceae. The family Betulaceae contains the genera _Betula_, birch; _Alnus_, alder; _Ostrya_, ironwood; and _Corylus_, hazel. These genera may be distinguished from each other by the general aspect of their pistillate inflorescences. That of _Betula_ is a simple ament. That of _Alnus_ is cone-like. That of _Ostrya_ has loosely associated involucral bracts. In _Ostrya virginiana_, the only species of our area, the common name "hop hornbeam" calls attention to an aspect resemblance of this inflorescence to that of the pistillate *Humulus lupulus*, hops, used in brewing to flavor beer. The pistillate inflorescence of *Corylus* is few-flowered and therefore not an ament, although it may be interpreted as phylogenetically reduced from a pistillate ament. The family Fagaceae contains the genera *Fagus*, beech; *Castanea*, chestnut; and *Quercus*, oak. These genera may be distinguished from each other by their fruits and associated involucral burs. That of *Fagus* is enclosed in a four-valved involucral bur. That of *Castanea* is enclosed in a two to four-valved prickly involucre. That of *Quercus* is only partially enclosed by the accrescent involucre, and the fruit is an acorn. Native species of all of these genera are cultivated together with introduced species of most of them. This is the most important angiosperm order for lumber trees, and many of its species are important as ornamentals. Some are important for their nuts. It is to be noted that all the Fagales genera here named have been important from ancient times and that in each case the generic name has been adopted from classical Latin. Latin *fagus* is cognate with English *beech*. Note the corresponding letters: f/b, a/ee, g/ch. This correspondence indicates the importance of the beech in times prior to classical antiquity.

Corylus is commonly included in the Betulaceae

(sometimes termed the Corylaceae). Eichler and Hutchinson recognize a Corylaceae separate from the Betulaceae. The Corylaceae thus delimited are not included in the glossary for Fagales.

14 2:1

BETULACEAE

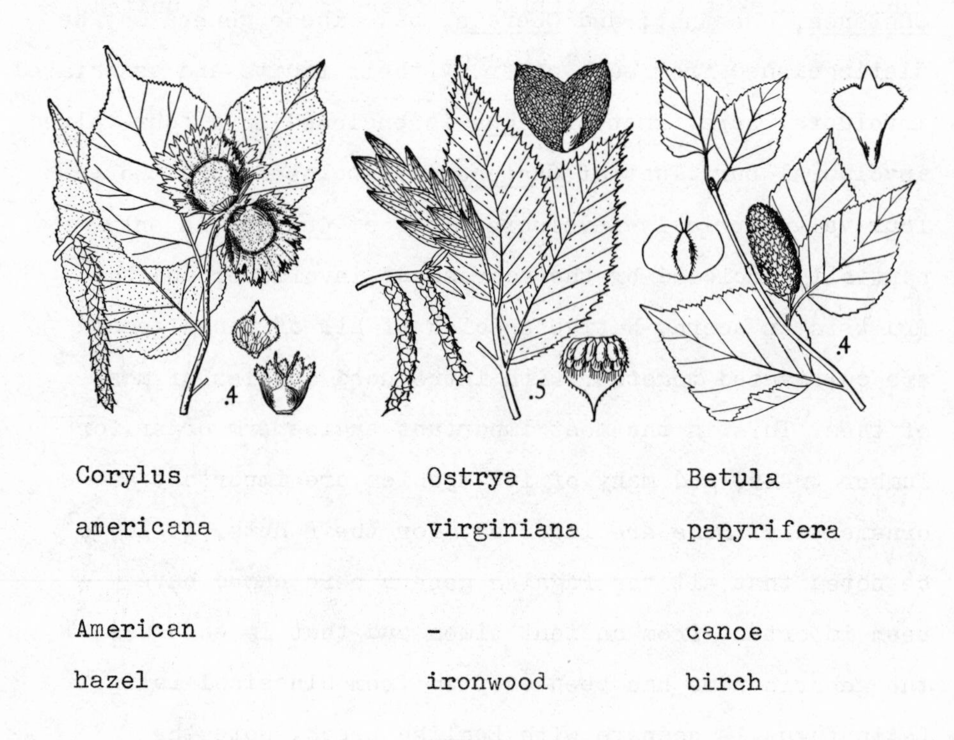

Corylus americana	Ostrya virginiana	Betula papyrifera
American hazel	- ironwood	canoe birch

BETULACEAE

dEpi,Psilo,G̸2L1A2-4(-10)P̸,i,4-0/monoec.nut ament.

Dicotyledonous plants with epigynous, or psilogynous,

flowers. Gynoecium of 2 united carpels and a single locule. Androecium of 2 to 4 stamens (or as many as 10). Perianth of from 4 tepals to none, united or free (when more than one). Plants monoecious. Fruit a nut. Inflorescences aments.

Betula sp.

dEpiG/2L1A2PO(f),/4(m)/monoec.

Perianth absent (in pistillate flowers), or of 4 united tepals (in staminate flowers).

FAGACEAE

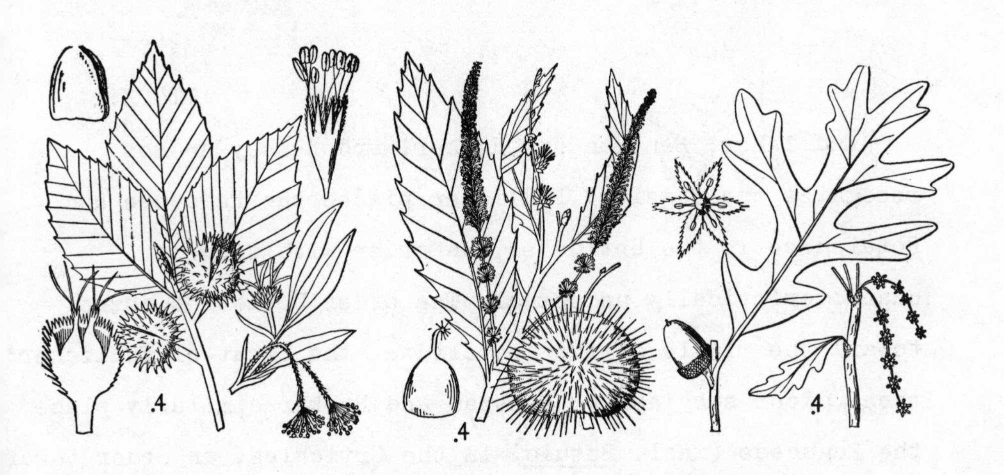

Fagus grandifolia	Castanea dentata	Quercus alba
- beech	- chestnut	white oak

FAGACEAE

dEpiG/3,(6),A3-nP/6(f),3-8(m),B(in frt.bur,cup)/monoec.nut ament.(m)

Dicotyledonous plants with epigynous flowers. Gynoecium of 3 united carpels (or in some cases 6). Androecium of 3 to many stamens. Perianth of united tepals; 6 (in pistillate flowers), or 3 to 8 (in staminate flowers). Bracts present (becomming in fruit a bur or cup). Plants monoecious. Fruit a nut. Inflorescence (of staminate flowers) an ament.

Quercus sp.

dEpiG/3A8P/6B(in frt.cup)/monoec.

14 3:1

Endlicher, Bentham and Hooker, and Bessey do not recognize the Fagales. Endlicher places the Fagaceae and Betulaceae in the Urticales, an order with flowers characteristically unisexual. The order Urticales comes toward the middle of the Apetalidae, the first of Endlicher's three dicot subclasses. Bentham and Hooker similarly place the Fagaceae (incl. Betula) in the Urticales, an order toward the end of their Apetalidae, the third of their three dicot subclasses. Eichler splits off the Fagaceae together with other families from the Urticales and raises the group to ordinal status based on ament inflorescences. Within this order Fagales Eichler includes not only the Fagaceae and

Betulaceae but also the Salicaceae, Myricaceae (incl. *Leitneria*), and Juglandaceae. Eichler places the Fagales toward the beginning of the Agamopetalidae, the second of his two dicot subclasses. Engler modifies Eichler's scheme by including only the Betulaceae and Fagaceae within the Fagales. Thus constituted, the Fagales differ from other amentiferous orders by having not only ovary inferior but also leaves simple. In the Engler system the orders derivative from Eichler's Fagales stand as a group toward the beginning of the Agamopetalidae, the first of Engler's two dicot subclasses. Bessey places the Betulaceae and Fagaceae, along with other amentiferous taxa, in the Sapindales. This order comes toward the middle of the Dicotocotyloididae, the second of his two dicot subclasses. Hutchinson places the Fagales in a hamamelidalian order group within the Lignosidae, the first of his two dicot subclasses. This group has characteristic inflorescences. Melchior places the Fagales in an apetalous order group within the Agamopetalidae, the first of his two dicot subclasses. He recognizes the order as highly advanced by reduction but casts doubt on any connection of it to the Hamamelidaceae. Cronquist, on the other hand, places the Fagales in the Hamamelidae, the second of his six dicot subclasses.

As already mentioned, Bentham and Hooker do not recognize the Betulaceae but include *Betula* and *Corylus* in the Fagaceae. Endlicher, Engler, Bessey, and Melchior agree on the sequence: Betulaceae, Fagaceae. In support of this position the nut fruits of the Fagaceae may be interpreted as advanced over the pistillate aments of the Betulaceae. Eichler and Hutchinson also accept the Betulaceae, Fagaceae sequence, but they split off the Corylaceae from the Betulaceae. Cronquist, on the other hand, prefers the sequence: Fagaceae, Betulaceae. He states that the Betulaceae are in some respects more advanced than the Fagaceae, and he casts doubt on any direct connection between the two families.

There is then a modern consensus that the Fagales are to be associated at least provisionally with other amentiferous orders. There is also consensus (Cronquist dissenting) that within the order the Betulaceae should precede the Fagaceae.

14 4:1

GLOSSARY FOR FAGALES

Fagales. Order in subclass Agamopetalidae.

Amentaceae. Unterreihe in Reihe Juliflorae (Eichl.)

In Klasse Chori- und Apetalae.

Fagales. Series in subclassis Archichlamydeae (Engl.)

Fagales. Order in division Lignosae (Hutch.)

Fagales. Reihe in Unterklasse Archichlamydeae (Melch.)

Fagales. Order in subclass Hamamelidae (Cronq.)

Betulaceae. Family in order Fagales.

Betulaceae. Ordo in classis Juliflorae (Endl.)

Betulaceae. Familie in Unterreihe Amentaceae (Eichl.)

Betulaceae. Familia in series Fagales (Engl.)

Betulaceae. Family in order Sapindales (Bessey).

Betulaceae. Family in order Fagales (Hutch.), (Cronq.)

Betulaceae. Familie in Reihe Fagales (Melch.)

Fagaceae. Family in order Fagales.

Cupuliferae. Ordo in classis Juliflorae (Endl.)

Cupuliferae. Ordo in series Unisexuales (B.&H.)

Cupuliferae. Familie in Unterreihe Amentaceae (Eichl.)

Fagaceae. Familia in series Fagales (Engl.)

Fagaceae. Family in order Sapindales (Bessey).

Fagaceae. Family in order Fagales (Hutch.), (Cronq.)

Fagaceae. Familie in Reihe Fagales (Melch.)

Salicaceae. See 10 4.

Salicineae. Familie in Unterreihe Amentaceae (Eichl.)

Myricaceae. See 11 4.

Myricaceae. Familie in Unterreihe Amentaceae (Eichl.)

Juglandaceae. See 13 4.

Juglandaceae. Familie in Unterreihe Amentaceae (Eichl.)

URTICALES

The order Urticales contains the families Ulmaceae, Moraceae, Cannabaceae, and Urticaceae. The family Ulmaceae usually contains the genus *Celtis*, hackberry, as well as *Ulmus*, elm. The flat samara fruits of *Ulmus* contrast with the spherical fruits of *Celtis*. The family Moraceae contains the genera *Morus*, mulberry, and *Maclura* Osage orange. The syncarp of *Morus* is juicy, but that of *Maclura* is covered by a dry rind. *Ficus*, fig, is an important genus of the family not represented in our area. The fruit of *Cannabis* occurs in axillary paniculate spikes. *Cannabis* commonly grows as a weed, but its name *C. sativa* L., "planted hemp," emphasizes its role as a cultivated plant and a source not only of hemp but also of such narcotic preparations as bhang, cannabin, charas, ganja, hashish, majoon, marijuana, etc. The fruit of *Humulus* occurs in druping paniculate clusters, hops, the glandular hairs of which impart flavor to beer. The family Urticaceae includes *Urtica*, nettle, and *Parietaria*, pellitory. *Urtica* has stinging hairs and opposite leaves. *Parietaria* lacks stinging hairs and has alternate leaves.

Family names for this order are based on *Ulmus*, *Morus*, *Cannabis*, and *Urtica*. All of them are ancient Latin forms, and *Cannabis* was adopted into Latin from Greek. It ultimately goes back to remoter times as evidenced by its cognate relationship with English *hemp*. Note the corresponding letters: c/h, nn/m, b/p. A similar cognate relationship may exist between Latin *ulmus* and English *elm*. The *mul*- of mulberry, on the other hand, is not cognate with the Latin *morus*, but simply derived from it.

The ancient Greek and Latin name *cannabis* was variously declined. The forms *Cannabaceae*, *Cannabiaceae*, and *Cannabidaceae* could all be based on ancient sources. Endlicher, however, used the family name *Cannab-ineae* (analogous to *Salic-ineae* based on *Salix*). Lindley (1847), in order to use the -*aceae* suffix dropped the final -*eae* from Endlicher's *Cannab-ineae* and added -*aceae*. This gives *Cannab-in-aceae*. *Cannabinaceae* became well-established in botanical usage, but it has recently been rejected. The family name is now formed on the stem *Cannab-i*- plus the suffix -*aceae*. It involves the loss of the *i* of the *i*-stem noun when followed by a suffix beginning with -*a*. For an objection to this change see *Taxon* 15: 205 seq. (1966).

Endlicher recognizes a family based on *Celtis* this is not included in the glossary for Urticales.

15 2:1

ULMACEAE

dHypoGi̸2L1A3-9Pi̸3-9/polyg.,monoec.,sam.,drup.

ULMACEAE

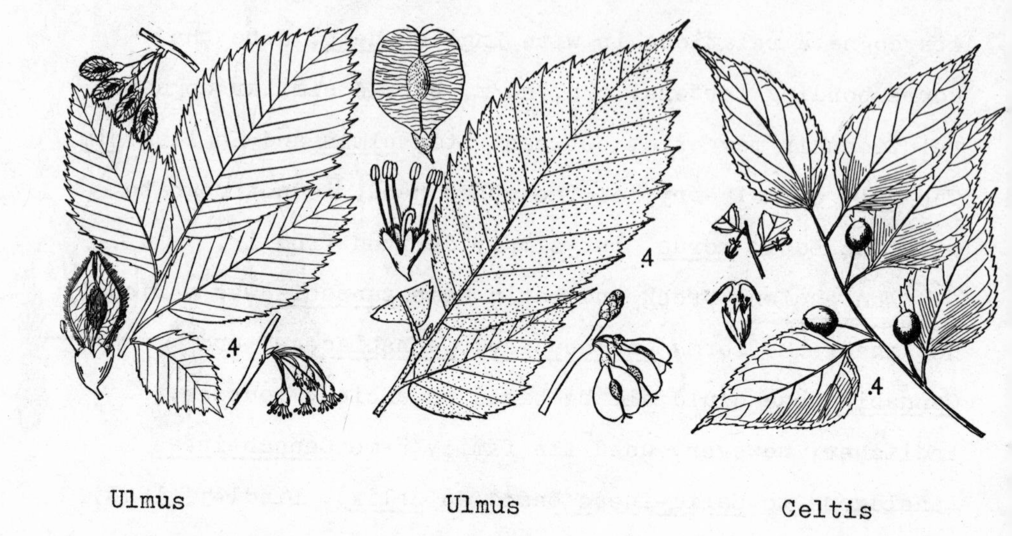

Ulmus americana	Ulmus rubra	Celtis occidentalis
American elm	slippery elm	hackberry

Dicotyledonous plants with hypogynous flowers. Gynoecium of 2 united carpels with a single locule. Androecium of 3 to 9 stamens. Perianth of 3 to 9 united tepals. Plants polygamous, or monoecious. Fruit a samara, or drupaceous.

Ulmus sp.

dHypoG/2L1A5P/5/polyg.,monoec.

MORACEAE

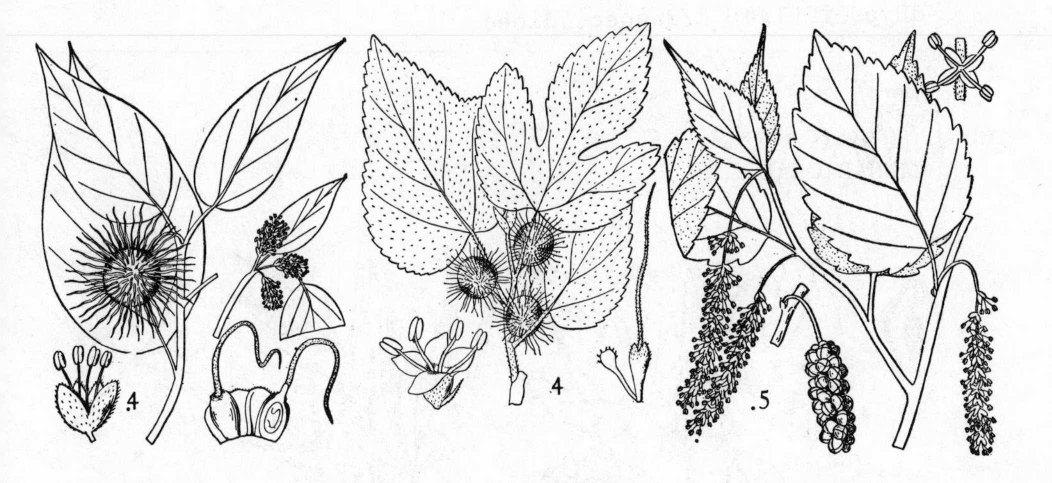

Maclura	Broussonetia	Morus
pomifera	papyrifera	rubra
Osage orange	paper	red
-	mulberry	mulberry

MORACEAE

dHypoGɨ2L̇1A4P_4/monoec.,dioec.,ach.(ɨ in multipl.frt.by fleshy,dry,P)

Dicotyledonous plants with hypogynous flowers. Gynoecium of 2 united carpels and a single locule. Androecium of 4 stamens. Perianth of 4 tepals, united at the base. Plants monoecious, or dioecious. Fruit an achene (united in a multiple fruit by a fleshy, or dry, perianth).

Morus sp.

dHypoG/2L1A4P_4/monoec.,dioec.

CANNABACEAE

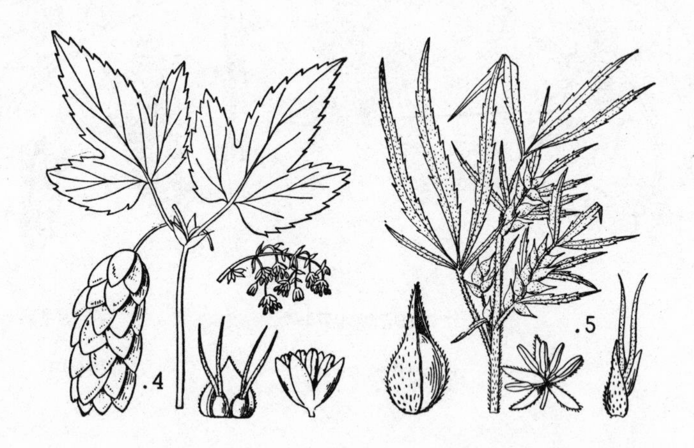

Humulus lupulus	Cannabis sativa
-	-
hop	hemp

CANNABACEAE

dHypoG/2L1A5P/ entire(f),5(m)/dioec.ach.

Dicotyledonous plants with hypogynous flowers. Gynoecium of 2 united carpels and a single locule. Androecium of 5 stamens. Perianth of united tepals, entire (in

pistillate flowers), or 5-lobed (in staminate flowers).
Plants dioecious. Fruit an achene.

Cannabis sativa

dHypoG1̸2L1A5P1̸ entire(f),5(m)/dioec.

URTICACEAE

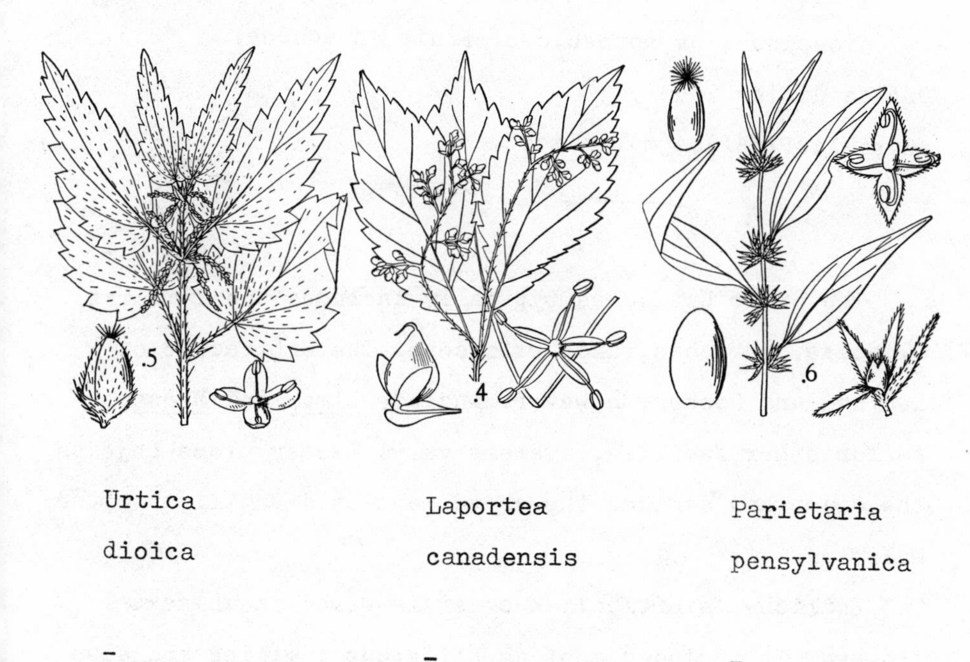

Urtica dioica	Laportea canadensis	Parietaria pensylvanica
-	-	-
nettle	wood nettle	pellitory

URTICACEAE

dHypoG1A(3-)4(-5)P1̸a̸(f),a(m,(f)),(3-)4(-5)/dioec.,monoec., ach.

Dicotyledonous plants with hypogynous flowers. Gynoecium of a single carpel. Androecium of 4 stamens (or as few as 3, or as many as 5). Perianth of 4 united tepals (or as few as 3, or as many as 5), zygomorphic (in the case of some pistillate flowers) or actinomorphic (in the case of staminate flowers (or in the case of some pistillate flowers)). Plants dioecious, or monoecious. Fruit an achene.

Urtica dioica

dHypoG1A4Pz/a(f),a(m),4/dioec.

15 3:1

The order Urticales typically includes the families: Ulmaceae, Moraceae, and Urticaceae. The Urticaceae of Bentham and Hooker, however, include _Ulmus_ and _Morus_. As for other families, systems vary. Bessey alone rejects the order and assigns the above mentioned families to the Malvales.

Endlicher's urticalian order is based on unisexual flowers. It includes most amentiferous families and also the Platanaceae. Within it he alone recognizes a family based on _Celtis_. The order is the third of his Apetalidae, the first of his three dicot subclasses. Bentham and Hooker's urticalian order is similarly inclusive. It too contains amentiferous families, but the group of these is not identical with Endlicher's group. The order further contains not only the Platanaceae but also the Euphorbiaceae. The order stands toward the end of the Apetalidae, the third of their three dicot subclasses. Eichler excludes the amentiferous families from his Urticales, but with diffidence he does admit the Platanaceae. The order stands in third place, after the Saururales and Fagales, in the apetalous portion of the Agamopetalidae, the second of his two dicot subclasses. Engler follows Eichler's basic pattern, but because of their minute petals he excludes the Platanaceae from the Urticales. His order stands in a position similar to that of Eichler's order, but numerically it is further back in its subclass because Engler has raised certain amentiferous families to ordinal rank. The first of Engler's two dicot subclasses is the Agamopetalidae. Bessey does not recognize the Urticales, and he places the Ulmaceae, Moraceae (incl. Cannabis), and Urticaceae in the Malvales. This is the second order of his Dicotostrobiloididae, the first of his two dicot subclasses. Hutchinson derives the order Urticales directly from the Hamamelidales. He therefore places the Urticales in a hamamelidalian order group within the Lignosidae, the first of his two dicot subclasses. This order group comes after magnolialian and rosalian order groups. Melchior retains the Urticales in the Englerian position within his admittedly artificial Agamopetalidae. Cronquist follows Hutchinson in deriving the Urticales from the Hamamelidales, and he makes Hutchinson's hamamelidalian order group into the Hamamelidae, the second of his six dicot subclasses.

The most common sequence of families corresponds to a reduction of the gynoecium within the order from dimeric to monomeric. This gives: Ulmaceae, Moraceae, Urticaceae—a sequence followed by most of the systems. Eichler alone reads the sequence in reverse order. The systems differ in their placement of _Cannabis_ or of the family Cannabaceae. Endlicher places the family after the Urticales. Bentham and Hooker include _Cannabis_ ahead of _Morus_ within their Urticaceae. Hutchinson adopts this sequence in placing the Cannabaceae ahead of the Moraceae within his Urticales. Eichler and Cronquist place the Cannabaceae after the Moraceae. Engler, Bessey, and Melchior include _Cannabis_ after _Morus_ within the Moraceae.

15 4:1

GLOSSARY FOR URTICALES

Urticales. Order in subclass Agamopetalidae.

Juliflorae. Classis in cohors Apetalae (Endl.)

Unisexuales. Series in subclassis Monochlamydeae (B.&H.)

Urticinae. Unterreihe in Reihe Juliflorae (Eichl.) In Klasse Chori- und Apetalae.

Urticales. Series in subclassis Archichlamydeae (Engl.)

Urticales. Order in division Lignosae (Hutch.)

Urticales. Reihe in Unterklasse Archichlamydeae (Melch.)

Urticales. Order in subclass Hamamelidae (Cronq.)

Ulmaceae. Family in order Urticales.

Ulmaceae. Ordo in classis Juliflorae (Endl.)

Ulmaceae. Familie in Unterreihe Urticinae (Eichl.)

Ulmaceae. Familia in series Urticales (Engl.)

Ulmaceae. Family in order Malvales (Bessey).

Ulmaceae. Family in order Urticales (Hutch.), (Cronq.)

Ulmaceae. Familie in Reihe Urticales (Melch.)

Moraceae. Family in order Urticales.

Moreae. Ordo in classis Juliflorae (Endl.)

Moraceae. Familie in Unterreihe Urticinae (Eichl.)

Moraceae. Familia in series Urticales (Engl.)

Moraceae. Family in order Malvales (Bessey).

Moraceae. Family in order Urticales (Hutch.), (Cronq.)

Moraceae. Familie in Reihe Urticales (Melch.)

Cannabaceae. Family in order Urticales.

Cannabineae. Ordo in classis Juliflorae (Endl.)

Cannabineae. Familie in Unterreihe Urticinae (Eichl.)

Cannabiaceae. Family in order Urticales (Hutch.)

Cannabaceae. Family in order Urticales (Cronq.)

Urticaceae. Family in order Urticales.

Urticaceae. Ordo in classis Juliflorae (Endl.)

Urticaceae. Ordo in series Unisexuales (B.&H.)

Urticaceae. Familie in Unterreihe Urticinae (Eichl.)

Urticaceae. Familia in series Urticales (Engl.)

Urticaceae. Family in order Malvales (Bessey).

Urticaceae. Family in order Urticales (Hutch.), (Cronq.)

Urticaceae. Familie in Reihe Urticales (Melch.)

Salicaceae. See 10 4.

Salicineae. Ordo in classis Juliflorae (Endl.)

Myricaceae. See 11 4.

Myriceae. Ordo in classis Juliflorae (Endl.)

Myricaceae. Ordo in series Unisexuales (B.&H.)

Leitneriaceae. See 12 4.

Leitnerieae. Ordo in series Unisexuales (B.&H.)

Juglandaceae. See 13 4.

Juglandeae. Ordo in series Unisexuales (B.&H.)

Betulaceae. See 14 4.

Betulaceae. Ordo in classis Juliflorae (Endl.)

Fagaceae. See 14 4.

Cupuliferae. Ordo in classis Juliflorae (Endl.)

Cupuliferae. Ordo in series Unisexuales (B.&H.)

Platanaceae. See 24 4.

Plataneae. Ordo in classis Juliflorae (Endl.)

Platanaceae. Ordo in series Unisexuales (B.&H.)

Platanaceae. Familie in Unterreihe Urticinae (Eichl.)

Euphorbiaceae. See 25 4.

Euphorbiaceae. Ordo in series Unisexuales (B.&H.)

SANTALALES

The order Santalales is only sparsely represented in our area, and it is not given regular coverage in the present survey. *Comandra umbellata* may be taken as a representative of the Santalaceae. *Comandra* is indicated by its name to have distinctive staminal hairs. It may also be given the name *false bastard-toadflax*. The term *false* distinguishes the genus from the santalaceous *Thesium*, bastard toadflax. The term *bastard* perhaps indicates not only that these taxa are not to be confused with the scrophulariaceous genus *Linaria*, toadflax, but that they have a reprehensible parasitic habit. The element *toad-* indicates not only that *Linaria* is not to be identified with *Linum*, flax, a distinction also expressed by the suffix -*aria*, but also that *Linaria* resembles toad's-mouth, *Antirrhinum*. *Comandra* is a parasite on roots of various plants.

Santalum is a generic name for sandal-wood. It is the late Latin name derived through Greek and Persian from Sanskrit *candana* (the *c* being now pronounced "tsh," not "k"). Sandal-wood is the source of a perfume used in rituals.

SANTALACEAE

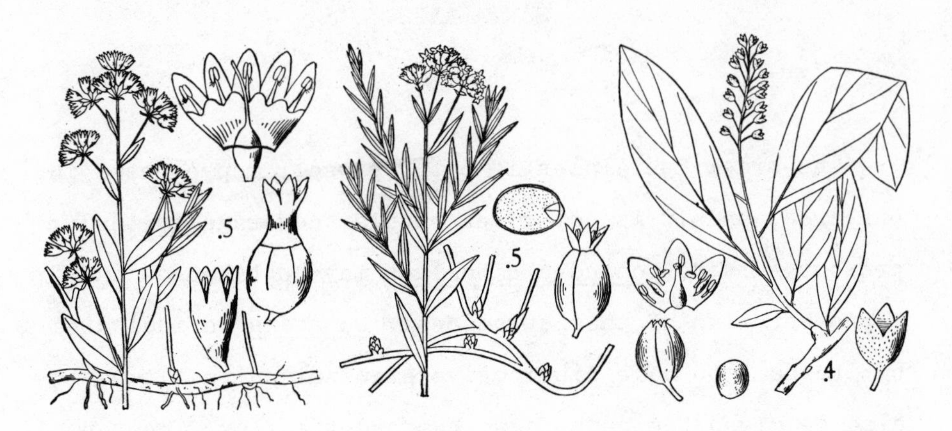

Comandra umbellata	Comandra pallida	Pyrularia pubera
bastard toad-flax	bastard toad-flax	- buffalo nut

SANTALACEAE

dEpi(w.disk)G/3-5L1A4-5(epitep.opp.)P4-5/perf.,monoec., polyg.drup.

Dicotyledonous plants with epigynous flowers having a disk. Gynoecium of 3 to 5 united carpels and a single locule. Androecium of 4 to 5 stamens epitepalous and opposite the tepals. Perianth of 4 to 5 tepals. Flowers perfect, or unisexual and the plants monoecious, or dioecious, or the plants polygamous. Fruit a drupe.

Comandra sp.

dEpi(w.disk)Gi̸3L1A5(epitep.opp.)P5/perf.

16 3:1

Endlicher does not recognize the Santalales. He assigns the Santalaceae to the Elaeagnales. Bentham and Hooker place the order Santalales in isolated position. Their name <u>Achlamydosporae</u> calls attention to the loss of seed coats in this parasitic group. Eichler's order Santalales is based on parasitic habit. So constituted it stands more or less outside his classification system. Engler associates the Santalales with the Aristolochiales and the Polygonales. Bessey does not recognize the Santalales. He assigns the Santalaceae to the Celastrales, an order which he derives from the Rosales. Hutchinson diagrams as a phylogenetic sequence: Celastrales, Santalales. Melchior maintains the order Santalales in its Englerian position. Cronquist diagrams as a phylogenetic sequence: Rosales, Santalales.

GLOSSARY FOR SANTALALES

Santalales. Order in subclass Agamopetalidae.

Achlamydosporeae. Series in subclassis Monochlamydeae (B.&H.)

Hysterophyta. Anhang in Klasse Chori- und Apetalae (Eichl.)

Santalales. Series in subclassis Archichlamydeae (Engl.)

Santalales. Order in division Lignosae (Hutch.)

Santalales. Reihe in Unterklasse Archichlamydeae (Melch.)

Santalales. Order in subclass Rosidae (Cronq.)

Santalaceae. Family in order Santalales.

Santalaceae. Ordo in classis Thymeleae (Endl.)

Santalaceae. Ordo in series Achlamydosporeae (B.&H.)

Santalaceae. Familie in Anhang Hysterophyta (Eichl.)

Santalaceae. Familia in series Santalales (Engl.)

Santalaceae. Family in order Celastrales (Bessey).

Santalaceae. Family in order Santalales (Hutch.), (Cronq.)

Santalaceae. Familie in Reihe Santalales (Melch.)

Aristolochiaceae. See 17 4.

Aristolochiaceae. Familie in Anhang Hysterophyta (Eichl.)

ARISTOLOCHIALES

The order Aristolochiales is represented in our area only by the family Aristolochiaceae. The order is not given regular coverage in the present survey. The only genera of the family in our area are Asarum and Aristolochia.

The name Aristolochia is derived through ancient Latin from Greek. It signifies: "plant for best childbirth," and the English name birthwort was coined as a translation for the Greek word. Aristolochia durior has the English name Dutchman's pipe descriptive of its s-shaped perianth.

17 2:1

ARISTOLOCHIACEAE

dEpiG/6A6,12,CO,3(minute);P,K,/a,ɑ/,3(tubular)/perf.caps.

Dicotyledonous plants with epigynous flowers. Gynoecium of 6 united carpels. Androecium of 6, or 12, stamens. Corolla none, or of 3 minute petals. Perianth, or calyx, of 3 united tepals, or sepals, (tubular), actinomorphic, or zygomorphic. Flowers perfect. Fruit a capsule.

Asarum sp.

dEpiG/6A12P/3/perf.

ARISTOLOCHIACEAE

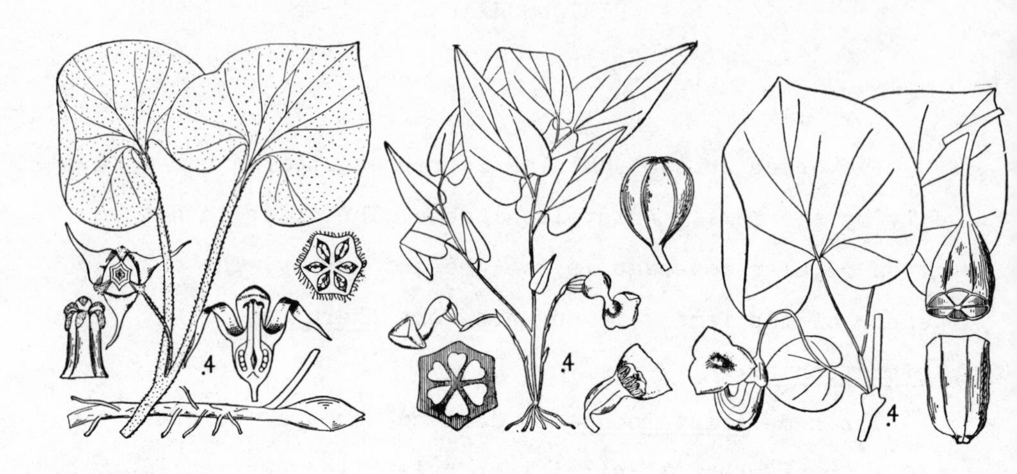

Asarum	Aristolochia	Aristolochia
canadense	serpentaria	durior
-	Virginia	Dutchman's pipe
wild ginger	snakeroot	-

17 3:1

Eichler and Bessey do not recognize the Aristolochiales. Eichler places the Aristolochiaceae in the Santalales. Bessey places the Aristolochiaceae in the Onagrales. Endlicher, Bentham and Hooker, and Engler classify the Aristolochiales with their apetalous orders. Hutchinson points to the petaliferous aristolochiacean genus Saruma as incompatible with the apetalous-group classification. Hutchinson diagrams as a phylogenetic sequence: Ranunculales,

Berberidales, Aristolochiales. Melchior associates the Aristolochiales with the ranalian complex. Cronquist derives the Aristolochiales directly from the Magnoliales.

17 4:1

GLOSSARY FOR ARISTOLOCHIALES

Aristolochiales. Order in subclass Agamopetalidae.
  Serpentariae. Classis in cohors Apetalae (Endl.)
  Multiovulatae terrestres. Series in subclassis Monochlamydeae (B.&H.)
  Aristolochiales. Series in subclassis Archichlamydeae (Engl.)
  Aristolochiales. Order in division Herbaceae (Hutch.)
  Aristolochiales. Reihe in Unterklasse Archichlamydeae (Melch.)
  Aristolochiales. Order in subclass Magnoliidae (Cronq.)

Aristolochiaceae. Family in order Aristolochiales.
  Aristolochieae. Ordo in classis Serpentariae (Endl.)
  Aristolochiaceae. Ordo in series Multiovulatae terrestres (B.&H.)
  Aristolochiaceae. Familie in Anhang Hysterophyta (Eichl.)
  Aristolochiaceae. Familia in series Aristolochiales (Engl.)
  Aristolochiaceae. Family in order Myrtales (Bessey).

Aristolochiaceae. Family in order Aristolochiales (Hutch.), (Cronq.)

Aristolochiaceae. Familie in Reihe Aristolochiales (Melch.)

POLYGONALES

The order Polygonales includes only the family Polygonaceae. Genera of the family occuring in our area include Rumex, dock; Polygonum, knotweed; and Fagopyrum, buckwheat. Characteristic of the family are ocreae, fused stipules forming sheaths around the stem at the leaf bases. Most species in the family are weedy herbs. Fagopyrum species have long been cultivated for their achene fruits. Beechwheat, an English name for the genus was in use as early as 1577, comparing the three-cornered achene to the beechnut. The modern English buckwheat is probably a dialectical version of beechwheat, and Fagopyrum is a translation of the English word into botanical Latin. Rheum rhaponticum, cultivated rhubarb, is also a member of this family. The Latin name means literally: "Volga-plant of the Volga river." The name rhubarb is derived from ancient names meaning literally: "Volga plant of the barbarian regions."

Polygonum is a Latinization of the Greek polygonon, knotgrass, etc. Literally the name means "multi-angled," in allusion to the many angular joints of the plant stem.

POLYGONACEAE

Rumex	Polygonum	Fagopyrum
crispus	pensylvanicum	esculentum
curled	smartweed	-
dock	-	buckwheat

POLYGONACEAE

dHypoGi̸2-3(-4)L1A6-9Pi,i̸,3-6/nut,ach.,ocr.

Dicotyledonous plants with hypogynous flowers. Gynoecium of 2 to 3 (or 4) united carpels and a single locule. Androecium of 6 to 9 stamens. Perianth of 3 to 6 tepals, separate or united. Fruit a nut, or achene. Leaves with ocreae.

Polygonum sp.

dHypoGi̸2L1A3+5Pi̸5/

Rumex sp.

dHypoGi̸3L1A3+5P5/

Fagopyrum sp.

dHypoGi̸3L1A6P6/

18 3:1

Endlicher, Bentham and Hooker, Eichler, and Bessey do not recognize the Polygonales. Endlicher and Bentham and Hooker place the Polygonaceae in their Chenopodiales. Eichler and Bessey place the Polygonaceae in their Caryophyllales, which include the Chenopodiaceae. Engler, Hutchinson, Melchior, and Cronquist accept the Polygonales with the Polygonaceae the only family of our area. Engler places the Polygonales toward the end of the apetalous orders within his Agamopetalidae, the first of his two dicot subclasses. Hutchinson places the Polygonales within the Herbacidae, the second of his two dicot subclasses. He diagrams as a phylogenetic sequence: Ranunculales, Caryophyllales, Polygonales. Melchior retains the

Polygonales in essentially their Englerian position. Cronquist places the Polygonales within the Caryophyllidae, the third of his six dicot subclasses. He diagrams as a phylogenetic sequence: Phytolaccaceae, Caryophyllaceae, Polygonaceae.

GLOSSARY FOR POLYGONALES

Polygonales. Order in subclass Agamopetalidae.

Polygonales. Series in subclassis Archichlamydeae (Engl.)

Polygonales. Order in division Herbaceae (Hutch.)

Polygonales. Reihe in Unterklasse Archichlamydeae (Melch.)

Polygonales. Order in subclass Caryophyllidae (Cronq.)

Polygonaceae. Family in order Polygonales.

Polygoneae. Ordo in classis Oleraceae (Endl.)

Polygonaceae. Ordo in series Curvembryeae (B.&H.)

Polygonaceae. Familie in Unterreihe Centrospermae (Eichl.)

Polygonaceae. Familia in series Polygonales (Engl.)

Polygonaceae. Family in order Caryophyllales (Bessey).

Polygonaceae. Family in order Polygonales (Hutch.), (Cronq.)

Polygonaceae. Familie in Reihe Polygonales (Melch.)

CARYOPHYLLALES A

The order Caryophyllales includes the Caryophyllaceae, Phytolaccaceae, Nyctaginaceae, and Chenopodiaceae. The order contains several other families not included in the present survey. The name Caryophyllaceae refers to the pink family, but it signifies etymologically: "clove family." Linnaeus gave the carnation the name Dianthus caryophyllus, which signifies "the Dianthus with clove odor." The name Dianthus is derived from ancient Greek diosanthos and means literally "flower of Zeus." The Greek name karyophyllos had been in use from ancient times and signified "nut of leaves," with reference to the dried flower buds which are the cloves of commerce. Caryophyllus was accordingly introduced into botanical nomenclature for the myrtacean clove genus. It was also introduced as a synonym for Dianthus. Both generic uses of Caryophyllus are now obsolete, but the family name Caryophyllaceae, proposed in 1828 in the Conspectus regni vegetabilis per gradus naturales evoluti of H. G. L. Reichenbach has not been thereby disqualified. The name Phytolaccaceae is derived from Phytolacca, which signifies literally "plant with crimson (berry) pigment." The name Nyctaginaceae is derived from Nyctago, and it signifies "night flower." The generic name is now obsolete. The plants it named are now usually merged into *Mirabilis*, which signifies "wonderful flower" presumably out of regard to its opening late in the day. The English name *four-o'clock* also refers to time of opening. The name *Chenopodiaceae* is derived from *Chenopodium*, and signifies: "plants with leaf shape resembling a goosefoot." The two components of *Chenopodium* and of *goosefoot* are cognate. Note corresponding letters as follows: (1) ch/g, n/-(*n* lost in goose, but present in *gander*), and (2) p/f, o/oo, d/t. The descriptive name for this order is *Centrospermae*, which refers to seed development at the center of a usually unilocular ovary.

The order Caryophyllales is divided by Fernald into apetalous and petaliferous portions. The introduction and discussion for all families are included in the treatment of the apetalous taxa (19a), but the Caryophyllaceae formulae and glossary are given separately (19b).

19a 2:1

CHENOPODIACEAE

dHypoG/2-3L1A2-5P/2-5/perf.,monoec.,dioec.,nut,ach.

Dicotyledonous plants with hypogynous flowers. Gynoecium of 2 to 3 united carpels with a single locule. Androecium of 2 to 5 stamens. Perianth of 2 to 5 united tepals. Flowers perfect, or unisexual and the plants monoecious, or dioecious. Fruit a nut, or an achene.

CHENOPODIACEAE

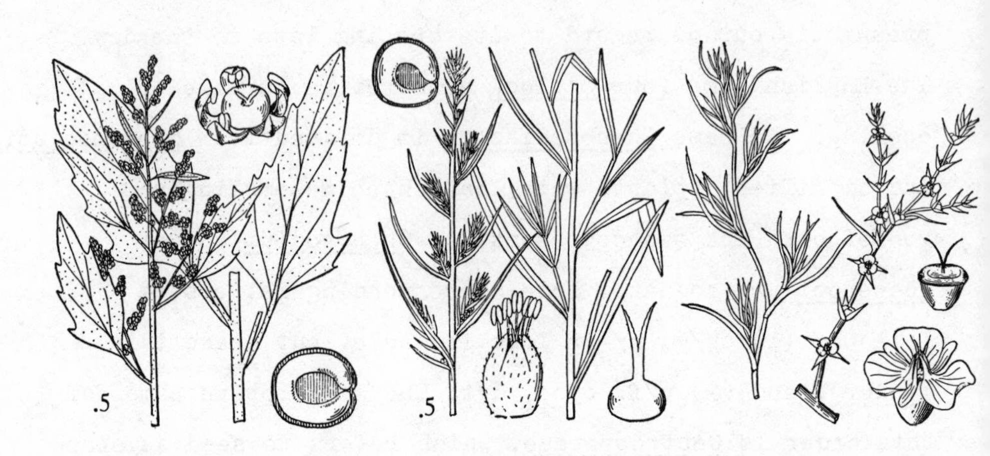

Chenopodium album

lamb's quarters

-

Kochia scoparia

summer cypress

-

Salsola kali var. tenuifoli

Russian thistl

-

Chenopodium album

dHypoGi̸2L1A5Pi̸5/perf.

NYCTAGINACEAE

dHypoG1Ai,_,3-5Pi̸5(corolloid)Bi̸5/perf.drup.,ach.

Dicotyledonous plants with hypogynous flowers. Gynoecium of a single carpel. Androecium of 3 to 5 stamens, free

NYCTAGINACEAE

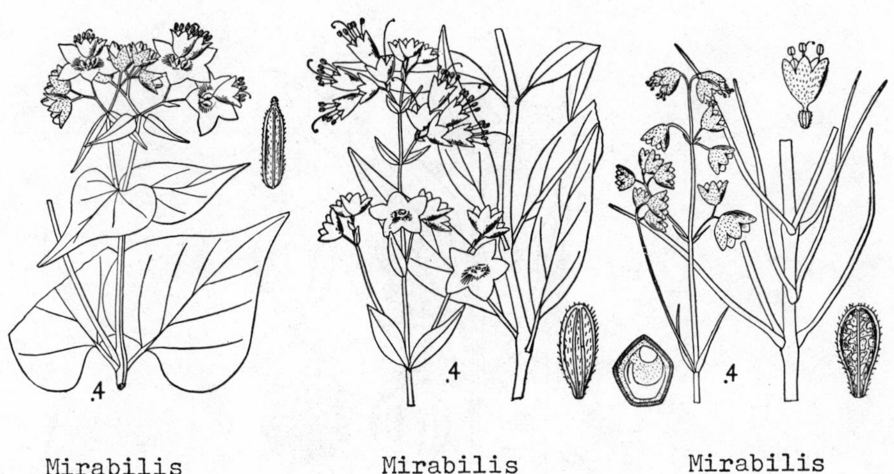

Mirabilis nyctaginea	Mirabilis albida	Mirabilis linearis
wild four-o'clock	-	-

or united at the base. Perianth of 5 united tepals (corolloid). Involucre of 5 united bracts. Flowers perfect. Fruit drupaceous, or an achene.

Mirabilis sp.

dHypoG1A5P/5(corolloid)B/5/perf.

PHYTOLACCACEAE

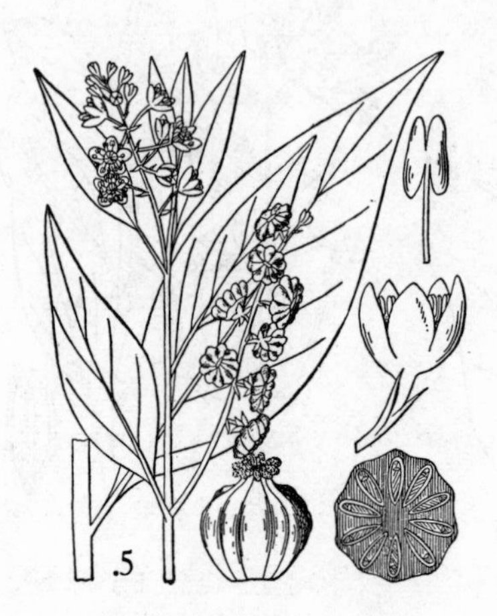

Phytolacca
americana

-

pokeweed

PHYTOLACCACEAE

dHypoG5̸-15A5-nP5/perf.ber.

Dicotyledonous plants with hypogynous flowers. Gynoecium of 5 to 15 united carpels. Androecium of 5 to many stamens. Perianth of 5 tepals. Flowers perfect. Fruit a berry.

Phytolacca americana

dHypoG1̸0A10P5/perf.

Endlicher recognizes both the Chenopodiales and Caryophyllales. He places the former at the end of the Apetalidae, the first of his three dicot subclasses. He places the latter in the middle of the Polypetalidae, the third of these subclasses. Bentham and Hooker's classification is similar, except that it places the Chenopodiales at the beginning of the Apetalidae and reverses the positions of the subclasses. Eichler includes the Chenopodiaceae in the order Caryophyllales at the end of his apetalous order group within the Agamopetalidae, the first of his two dicot subclasses. He explicitly notes the petaliferous caryophyllalian families as a transition from the apetalous to petaliferous condition. Engler follows Eichler's basic classification of the Caryophyllales. Bessey, however, reverses Eichler's position by finding in the apetalous caryophyllalian families a transition from the petaliferous to apetalous condition. Bessey places the Caryophyllales at the end of the polypetalous order group within the Dicotostrobiloididae, the first of his two dicot subclasses. Hutchinson not only separates the Chenopodiaceae from the Caryophyllales but he diagrams as a phylogenetic sequence: Ranunculales, Caryophyllales, Polygonales, Chenopodiales. He places these orders within the Herbacidae, the second of his two dicot subclasses. Melchior, while following the Englerian sequence in his placement of the Caryophyllales, adopts the Phytolaccaceae as the primitive family of the order. This family is characterized by a multilocular ovary, and it includes woody members and petaliferous members. Cronquist places the Caryophyllales within the Caryophyllidae, the third of his six dicot subclasses. This is a predominantly herbaceous subclass, and he derives all members of it directly or indirectly from the Phytolaccaceae.

Endlicher's order Chenopodiales runs: Chenopodiaceae, Polygonaceae, Nyctaginaceae. His order Caryophyllales runs: Caryophyllaceae, Phytolaccaceae. His family Phytolaccaceae is thus basically petaliferous. Bentham and Hooker's order Chenopodiales has the sequence: Nyctaginaceae, Chenopodiaceae, Phytolaccaceae, Polygonaceae. Their order Caryophyllales contains the Caryophyllaceae. Eichler adopts the sequence: Polygonaceae, Chenopodiaceae, Phytolaccaceae, Nyctaginaceae, Caryophyllaceae. Engler adopts Eichler's classification of the order, except that he excludes the Polygonaceae and transposes the Nyctaginaceae and Phytolaccaceae. Bessey's sequence of families is more or less the reverse of Engler's. It runs: Caryophyllaceae, Salicaceae, Podostemaceae, Phytolaccaceae, Chenopodiaceae, Polygonaceae, Nyctaginaceae. Hutchinson distributes these families into several orders. In his Herbacidae these are: Caryophyllales, Polygonales, and Chenopodiales (incl. Phytolaccaceae). In his Lignosidae the Onagrales include the Nyctaginaceae. Melchior's order

Caryophyllales runs: Phytolaccaceae, Nyctaginaceae, Chenopodiaceae, Caryophyllaceae. He derives the Nyctaginaceae, Chenopodiaceae, and Caryophyllaceae from the Phytolaccaceae by separate lines. Cronquist similarly derives these families from the Phytolaccaceae. In a parallel manner he derives the Cactaceae from the Phytolaccaceae, and he therefore includes the Cactaceae in the Caryophyllales.

GLOSSARY FOR CARYOPHYLLALES A

See also 19b 4.

Caryophyllales A. Order in subclass Agamopetalidae.

  Oleraceae. Classis in cohors Apetalae (Endl.)

  Curvembryeae. Series in subclassis Monochlamydeae (B.&H.)

  Centrospermae. Unterreihe in Reihe Centrospermae (sic) (Eichl.) In Klasse Chori- und Apetalae.

  Centrospermae. Series in subclassis Archichlamydeae (Engl.)

  Chenopodiales. Order in division Herbaceae (Hutch.)

  Centrospermae. Reihe in Unterklasse Archichlamydeae (Melch.)

Chenopodiaceae. Family in order Caryophyllales A.

  Chenopodeae. Ordo in classis Oleraceae (Endl.)

  Chenopodiaceae. Ordo in series Curvembryeae (B.&H.)

Chenopodiaceae. Familie in Unterreihe Centrospermae (Eichl

Chenopodiaceae. Familia in series Centrospermae (Engl.)

Chenopodiaceae. Family in order Caryophyllales (Bessey), (Cronq.)

Chenopodiaceae. Family in order Chenopodiales (Hutch.)

Chenopodiaceae. Familie in Reihe Centrospermae (Melch.)

Nyctaginaceae. Family in order Caryophyllales A.

Nyctagineae. Ordo in classis Oleraceae (Endl.)

Nyctagineae. Ordo in series Curvembryeae (B.&H.)

Nyctaginiaceae. Familie in Unterreihe Centrospermae (Eichl.)

Nyctaginaceae. Familia in series Centrospermae (Engl.)

Nyctaginaceae. Family in order Caryophyllales (Bessey), (Cronq.)

Nyctaginaceae. Family in order Thymelaeales (Hutch.)

Nyctaginaceae. Familie in Reihe Centrospermae (Melch.)

Phytolaccaceae. Family in order Caryophyllales A.

Phytolaccaceae. Ordo in classis Caryophyllinae (Endl.)

Phytolaccaceae. Ordo in series Curvembryeae (B.&H.)

Phytolaccaceae. Familie in Unterreihe Centrospermae (Eichl.)

Phytolaccaceae. Familia in series Centrospermae (Engl.)

Phytolaccaceae. Family in order Caryophyllales (Bessey), (Cronq.)

Phytolaccaceae. Family in order Chenopodiales (Hutch.)

Phytolaccaceae. Familie in Reihe Centrospermae (Melch.)

Polygonaceae. See 18 4.

Polygoneae. Ordo in classis Oleraceae (Endl.)

Polygonaceae. Ordo in series Curvembryeae (B.&H.)

Polygonaceae. Familie in Unterreihe Centrospermae (Eichl.)

POLYPETALOUS ORDER GROUPS

The orders of Endlicher's Polypetalidae run as follows: Umbellales, Saxifragales, Magnoliales, Papaverales, Nymphaeales, Violales, Cucurbitales, Cactales, Caryophyllales, Malvales, Theales, Sapindales, Celastrales, Euphorbiales, Rutales, Geraniales, Onagrales, Rosales, Leguminales. The sequence may be divided into five order groups as follows: (1) Umbellales, Saxifragales, (2) Magnoliales, Papaverales, Nymphaeales, (3) Violales, Cucurbitales, Cactales, Caryophyllales, Malvales, Theales, (4) Sapindales, Celastrales, Euphorbiales, Rutales, Geraniales, (5) Onagrales Rosales, Leguminales.

The orders of Bentham and Hooker's Polypetalidae are distributed into three superorders. The superorder Thalamiflorae contains orders which run: Magnoliales, Violales, Caryophyllales, Theales, Malvales. These may be divided into two order groups as follows: (1) Magnoliales, (2) Violales, Caryophyllales, Theales, Malvales. The orders of the superorder Disciflorae may be treated as a single order group: Geraniales, Aquifoliales, Celastrales, Sapindales. The orders of the superorder Calyciflorae may be treated as a single order group: Rosales, Onagrales, Cucurbitales, Cactales, Umbellales.

The orders of Eichler's polypetalous Agamopetalidae are distributed into four superorders. The superorder Aphanocyclicae contains orders which run: Magnoliales, Papaverales, Theales, Malvales. These may be divided into two order groups as follows: (1) Magnoliales, Papaverales, (2) Theales, Malvales. The orders of the superorder Eucyclicae may be treated as a single order group: Geraniales, Rutales, Sapindales, Celastrales. The superorder Tricoccae contains a single order: Euphorbiales. The superorder Calyciflorae contains orders which run: Umbellales, Saxifragales, Onagrales, Elaeagnales, Rosales, Leguminales. These may be divided into two order groups as follows: (1) Umbellales, Saxifragales, (2) Onagrales, Elaeagnales, Rosales, Leguminales.

Engler makes a major distinction between the hypogynous-and-perigynous polypetalous Agamopetalidae and the epigynous polypetalous Agamopetalidae, although he recognizes the transition between these groups to be gradual. The orders of Engler's hypogynous-and-perigynous polypetalous Agamopetalidae run as follows: Magnoliales, Papaverales, Rosales, Geraniales, Sapindales, Rhamnales, Malvales, Theales. These may be divided into four order groups as follows: (1) Magnoliales, Papaverales, (2) Rosales, (3) Geraniales, Sapindales, Rhamnales, (4) Malvales, Theales. The Englerian basically epigynous polypetalous Agamopetalidae are the Cactales, Onagrales, and Umbellales.

The dicotyledons are divided by Bessey into two subclasses, the Dicotostrobiloididae and the Dicotocotyloididae, based primarily on hypogynous as opposed to perigynous or epigynous flowers. Each of these subclasses is divided into polypetalous and gamopetalous primary groups. Bessey does not have a separate apetalous group but distributes the apetalous families among putatively related polypetalous orders. The orders of Bessey's polypetalous Dicotostrobiloididae are included in a superorder with the following orders: Magnoliales, Malvales, Geraniales, Theales, Papaverales, Caryophyllales. These may be divided into three order groups as follows: (1) Magnoliales, Malvales, (2) Geraniales, (3) Theales, Papaverales, Caryophyllales. His polypetalous Dicotocotyloididae are included in a superorder with the following orders: Rosales, Onagrales, Cucurbitales, Cactales, Celastrales, Sapindales, Umbellales. These may be divided into two order groups as follows: (1) Rosales, Onagrales, Cucurbitales, Cactales, (2) Celastrales, Sapindales, Umbellales.

The dicotyledons are divided by Hutchinson into two subclasses, the Lignosidae and Herbacidae. His Lignosidae have the sequence: Magnoliales, Annonales, Rosales, Leguminales, Cornales, Hamamelidales, Salicales, Fagales, Juglandales, Urticales, Violales, Cucurbitales, Cactales, Tiliales, Malvales, Euphorbiales, Theales, Ericales,

Celastrales, Rhamnales, Ebenales, Rutales, Sapindales, Apocynales. The Hamamelidales, Salicales, Fagales, Juglandales, and Urticales are mentioned under 9 Oc. The remaining orders may be divided into order groups as follows: (1) Magnoliales, Annonales, (2) Rosales, Leguminales, Cornales, (3) Violales, Cucurbitales, Cactales, Tiliales, Malvales, (4) Euphorbiales, (5) Theales, Ericales, (6) Celastrales, Rhamnales, Ebenales, Rutales, Sapindales, Apocynales. Note the placement here of the Rutales, remote from the Geraniales associated by Hutchinson with the Polemoniales. The agamopetalous Herbacidae of Hutchinson have the sequence: Ranunculales, Berberidales, Papaverales, Cruciferales, Resedales, Caryophyllales, Polygonales, Chenopodiales, Onagrales, Gentianales, Primulales, Plantaginales, Saxifragales, Umbellales. The sequence may be divided into order groups as follows: (1) Ranunculales, Berberidales, Papaverales, Cruciferales, Resedales, (2) Caryophyllales, Polygonales, Chenopodiales, Onagrales, Gentianales, Primulales, Plantaginales, (3) Saxifragales, Umbellales.

The orders of Melchior's basically hypogynous polypetalous Agamopetalidae run as follows: Magnoliales, Ranunculales, Theales, Papaverales, Rosales, Geraniales, Rutales, Sapindales, Celastrales, Rhamnales, Malvales, Elaeagnales, Violales, Cucurbitales. The sequence may be divided into order groups analogous to those of the Englerian system as follows: (1) Magnoliales, Ranunculales,

Theales, Papaverales, (2) Rosales, (3) Geraniales, Rutales, Sapindales, Celastrales, Rhamnales, (4) Malvales, Elaeagnales, Violales, Cucurbitales. The basically epigynous polypetalous Agamopetalidae of Melchior are the Onagrales and Umbellales.

Cronquist recognizes magnolialian, hamamelidalian, caryophyllalian, thealian, and rosalian basically agamopetalous subclasses. His Hamamelidae are mentioned under 9 Oc. The orders of his Magnoliidae run: Magnoliales, Nymphaeales, Ranunculales, Papaverales. His Caryophyllidae run: Caryophyllales, Polygonales. His Theidae are arranged in two order groups. The first of these is polypetalous, and its orders run: Theales, Malvales, Violales, Salicales, Capparales. His Rosidae run: Rosales, Onagrales, Elaeagnales, Cornales, Celastrales, Euphorbiales, Rhamnales, Sapindales, Geraniales, Linales, Umbellales.

19b Ob:1

GLOSSARY FOR POLYPETALOUS SUPERORDERS, ETC.

Thalamiflorae (cohortes: Ranales, Parietales, Caryophyllinae, Guttiferales, Malvales). Series in subclassis Polypetalae (B.&H.)

Disciflorae (cohortes: Geraniales, Olacales, Celastrales, Sapindales). Series in subclassis Polypetalae (B.&H.)

Calyciflorae (cohortes: Rosales, Myrtales, Passiflorales, Ficoidales, Umbellales). Series in subclassis Polypetalae (B.&H.)

Aphanocyclicae (Unterreihen: Polycarpicae, Rhoeadinae, Cistiflorae, Columniferae). Reihe in Klasse Chori- und Apetalae (Eichl.)

Eucyclicae (Unterreihen: Gruinales, Terebinthinae, Aesculinae, Frangulinae). Reihe in Klasse Chori- und Apetalae (Eichl.)

Tricoccae (Unterreihe: Tricoccae (sic) (Euphorbiaceae, etc.)) Reihe in Klasse Chori- und Apetalae (Eichl.)

Calyciflorae (Unterreihen: Umbelliflorae, Saxifraginae, Myrtiflorae, Thymelaeinae, Rosiflorae, Leguminosae, (Anhang) Histerophyta). Reihe in Klasse Chori- und Apetalae (Eichl.)

Apopetalae-Polycarpellatae (orders: Ranales, Malvales, Sarraceniales, Geraniales, Guttiferales, Rhoeadales, Caryophyllales). Superorder in subclass Strobiloideae (Bessey).

Apopetalae (orders: Rosales, Myrtales, Loasales, Cactales, Celastrales, Sapindales, Umbellales). Superorder in subclass Cotyloideae (Bessey).

CARYOPHYLLALES B

The order Caryophyllales has already been considered under 19a 1. In Fernald's classification, however, the order is divided into an a (apetalous) part and a b (polypetalous) part. The Caryophyllaceae are included in this polypetalous part.

19b 2:1

CARYOPHYLLACEAE

dHypoGi̸2-5L1A(2-)8-10C(3-)4-5Ki,_,(3-)4-5/caps./plac. free-centr.

Dicotyledonous plants with hypogynous flowers. Gynoecium of 2 to 5 united carpels and a single locule. Androecium of 8 to 10 stamens (or as few as 2). Corolla of 4 to 5 petals (or as few as 3). Calyx of 4 to 5 sepals (or as few as 3), free or united at the base. Fruit a capsule. Placentation free-central.

Stellaria sp.

dHypoGi̸3L1A10C5K5/

CARYOPHYLLACEAE

Stellaria media	Saponaria officinalis	Silene nivea
common chickweed	bouncing bet –	white campion

19b 3:1

Of the several petaliferous caryophyllalian families only the Caryophyllaceae are included in the present survey. Discussion of Caryophyllaceae classification in the eight systems here considered is included under 19a 3.

GLOSSARY FOR CARYOPHYLLALES B

See also 19a 4.

Caryophyllales B. Order in subclass Agamopetalidae.

Caryophyllinae. Classis in cohors Dialypetalae (Endl.)

Caryophyllinae. Cohors in series Thalamiflorae (B.&H.) In subclassis Polypetalae.

Caryophyllales. Order in superorder Apopetalae-Polycarpellatae (Bessey). In subclass Strobiloididae.

Caryophyllales. Order in division Herbaceae (Hutch.)

Caryophyllales. Order in subclass Caryophyllidae (Cronq.)

Caryophyllaceae. Family in order Caryophyllales B.

Caryophylleae. Ordo in classis Caryophyllinae (Endl.)

Caryophylleae. Ordo in cohors Caryophyllinae (B.&H.)

Caryophyllaceae. Familie in Unterreihe Centrospermae (Eichl.)

Caryophyllaceae. Familia in series Centrospermae (Engl.)

Caryophyllaceae. Family in order Caryophyllales (Bessey), (Hutch.), (Cronq.)

Caryophyllaceae. Familie in Reihe Centrospermae (Melch.)

Salicaceae. See 10 4.

Salicaceae. Family in order Caryophyllales (Bessey).

Polygonaceae. See 18 4.

Polygonaceae. Family in order Caryophyllales (Bessey).

Podostemaceae. See 23 4.

Podostemonaceae. Family in order Caryophyllales (Bessey).

Cactaceae. See 30 4.

Cactaceae. Family in order Caryophyllales (Cronq.)

MAGNOLIALES

The order Magnoliales includes the Magnoliaceae, Annonaceae, Ranunculaceae, Nymphaeaceae, and Berberidaceae. The type genera of these families are included within our area except in the case of the Annonaceae which are represented by Asimina.

The Magnoliaceae on account of their numerous spirally-arranged floral parts and woody structure are commonly regarded as the most primitive angiosperm family. The Annonaceae resemble the Magnoliaceae in several respects, but in the Annonaceae the petals are six in two ranks. Asimina has elipsoid edible fruits. The Ranunculaceae are herbaceous and display transition in floral structure from a type resembling the Magnoliaceae to the definite numbers of floral parts found in many higher families. The Nymphaeaceae are aquatics which display resemblances to alismatalian monocots, and a close connection between the two groups has often been suggested. The Berberidaceae are characterized by hexamerous whorls usually interpreted as composed of two trimerous whorls each. The trimerous character has been given considerable weight by Hallier, who sees in the Berberidaceae the primitive angiosperm type. In the dicots the loss of one member in the hexamerous whorls is supposed to result in the common pentamerous condition, and in the monocots loss of trimerous whorls is supposed to result in a residual trimerous condition.

Nelumbo and Cabomba are commonly included in the Nymphaeaceae. Endlicher and Bessey recognize both a Nelumbonaceae and a Cabombaceae. Hutchinson recognizes a Cabombaceae and Cronquist recognizes a Nelumbonaceae. The Nelumbonaceae and Cabombaceae are not included in the following discussion, and the names Cambombeae (Endl.), Cambombaceae (Bessey), (Hutch.), Nelumboneae (Endl.), Nelumbaceae (Bessey), and Nelumbonaceae (Cronq.) are not included in the glossary for Magnoliales.

Podophyllum is commonly included in the Berberidaceae, but Cronquist puts it in the Ranunculaceae. Hutchinson recognizes a family Podophyllaceae. This family is not included in the following discussion, and the name Podophyllaceae is not included in the glossary for Magnoliales.

Nymphaea is the ancient Latin name for a water lily. The name was adopted from Greek usage which applied it to various water lilies. Its signification is: "plant with a flower suggestive of a water nymph." *Ranunculus* is an ancient Latin diminutive of *Rana*, "frog." It was also applied to crowfoot having a marshy habitat. This usage represents a loan translation of Greek *batrachion*, a derivative from *batrachos*, "frog." *Berberis* is a middle Latin word taken from vernacular Arabic. It is derived from standard Arabic *barbāris* and signifies: plant found in Barbary." The Arabic name for "Barbary" appears to be a specific application of the Greek *barbaros*, "barbarous, foreign," to a region. *Magnolia* is a name given by Plumier to commemorate the French botanist Pierre Magnol (1638-1715). *Annona* is a Latinized version of a Central American Taino Indian name for the custard-apple.

20 2:1

NYMPHAEACEAE

dHypoGi̸,i,2-nA3-nP5-n,C3-nK3-6/caps.,fol.,etc.,plac.par., marg./aquat.

Dicotyledonous plants with hypogynous flowers. Gynoecium of 2 to many united, or free, carpels. Androecium of 3 to many stamens. Perianth of 5 to many tepals, or a corolla of 3 to many petals and a calyx of 3 to 6 sepals. Fruit a capsule, or a follicle, etc. Placentation parietal, or marginal. Plants aquatic.

Nelumbo sp.

dHypoGnAnCnK6/

NYMPHAEACEAE

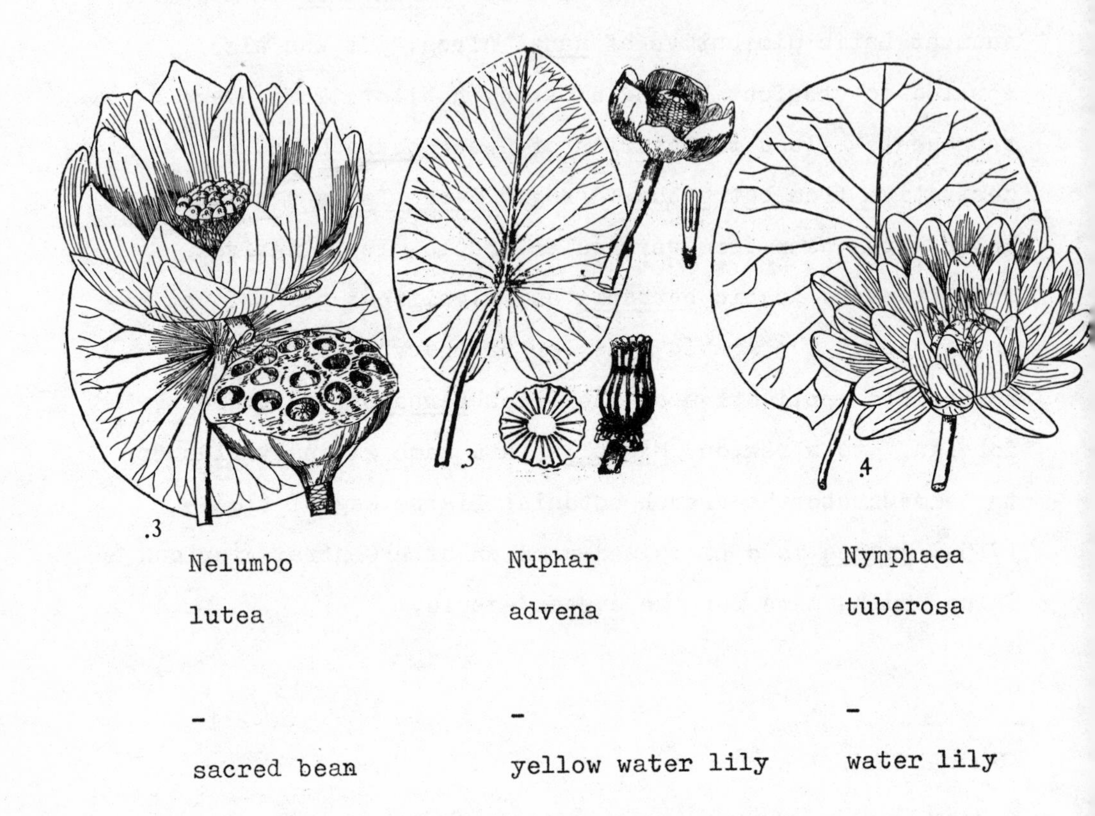

Nelumbo lutea	Nuphar advena	Nymphaea tuberosa
-	-	-
sacred bean	yellow water lily	water lily

RANUNCULACEAE

dHypoGl-nA(5-)nP3-n,Ca,ø,4-10K3-5/ach.,fol.,bac.,plac. marg.terr.,aquat.

Dicotyledonous plants with hypogynous flowers. Gynoecium of 1 to many carpels. Androecium of many stamens (or as few as 5). Perianth of 3 to many tepals, or corolla of 4 to 10 petals, actinomorphic or zygo-

RANUNCULACEAE

Ranunculus	Delphinium	Myosurus
acris	virescens	minimus
-	-	-
buttercup	larkspur	mousetail

morphic, and the calyx of 3 to 5 sepals. Fruit an achene, or follicle, or baccate. Placentation marginal. Plants terrestrial, or aquatic.

Ranunculus sp.

dHypoGnAnC5K5/

BERBERIDACEAE

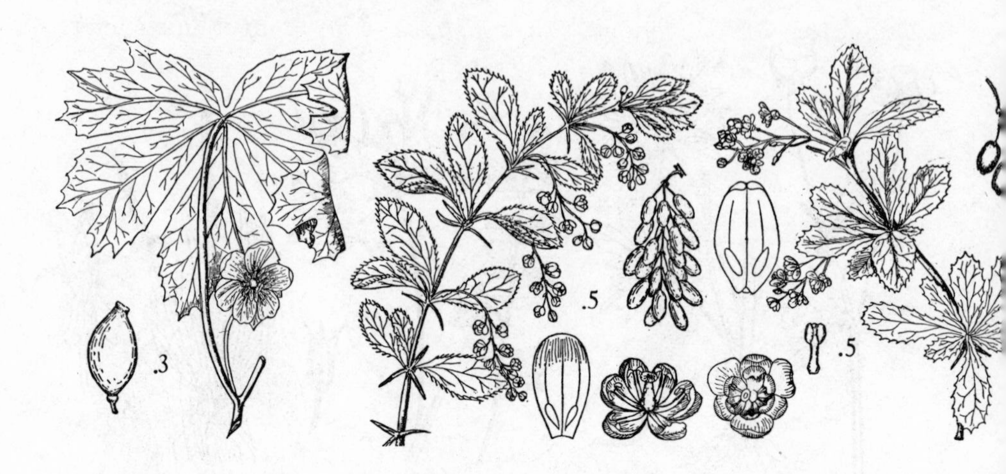

Podophyllum peltatum	Berberis vulgaris	Berberis canadensis
- May apple	European barberry	American barberry

BERBERIDACEAE

dHypoGlA6(-n)C6(-9)K(4-)6/ber.,caps.

Dicotyledonous plants with hypogynous flowers. Gynoecium of a single carpel. Androecium of 6 stamens (or as many as more than 10). Corolla of 6 petals (or as many as 9). Calyx of 6 sepals (or as few as 4). Fruit a berry, or capsule.

Berberis sp.

dHypoGlA6(3+3)C6(3+3)K6(3+3)/

Androecium of 6 stamens (in two cycles of 3 stamens each).

MAGNOLIACEAE

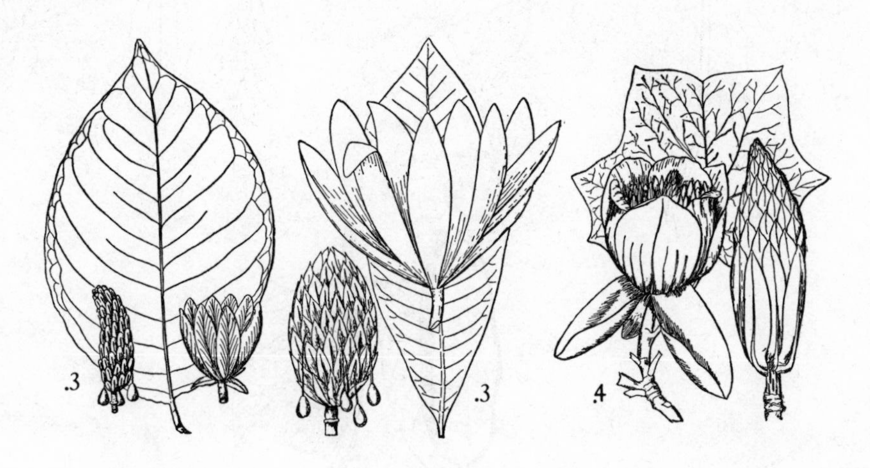

Magnolia	Magnolia	Liriodendron
acuminata	tripetala	tulipifera
-	-	-
		tulip tree

MAGNOLIACEAE

dHypoGnAnP9-15,C6K3/fol.,sam./arom.

Dicotyledonous plants with hypogynous flowers. Gynoecium of numerous carpels. Androecium of numerous stamens. Perianth of 9 to 15 tepals, or corolla of 6 petals and calyx of 3 sepals. Fruit a follicle, or samara. Plants aromatic.

Liriodendron tulipifera

dHypoGnAnC6K3/

ANNONACEAE

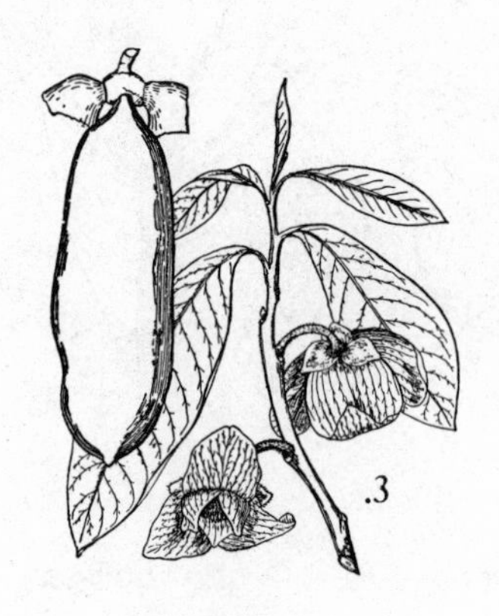

Asimina

triloba

-

pawpaw

ANNONACEAE

dHypoG3-15AnC6K3/large ber./arom.

Dicotyledonous plants with hypogynous flowers. Gynoecium

of 3 to 15 carpels. Androecium of numerous stamens. Corolla of 6 petals. Calyx of 3 sepals. Fruit a large berry. Plants aromatic.

Asimina triloba

dHypoG6AnC6K3/

20 3:1

Endlicher recognizes both the Magnoliales and the Nymphaeales. He places them toward the beginning of the Polypetalidae, the third of his three dicot subclasses. Bentham and Hooker place the Magnoliales in first position in the superorder Thalamiflorae within the Polypetalidae, the first of their three dicot subclasses. Eichler places the Magnoliales at the beginning of his non-eucyclic superorder Aphanocyclicae within the Agamopetalidae, the second of his two dicot subclasses. Engler discards Eichler's superorder, but he follows Eichler's general sequence of orders in his placement of the Magnoliales at the beginning of the basically polypetalous portion of the subclass Agamopetalidae, the first of his two dicot subclasses. Bessey places the Magnoliales in first position in the superorder Apopetalae-Polycarpellatae within the Dicotostrobiloididae, the first of his two dicot subclasses. Hutchinson recognizes the Magnoliales and Annonales and places them at the beginning of the Lignosidae, the first of his two dicot subclasses. He also recognizes the

Ranunculales and Berberidales and places them at the beginning of the Herbacidae, the second of his subclasses. Melchior also accepts both the Magnoliales and Ranunculales, but he places them together to jointly correspond to Engler's more broadly circumscribed Magnoliales. Cronquist recognizes the Magnoliales, Nymphaeales, and Ranunculales; and he places the three orders in the Magnoliidae, the first of his six dicot subclasses. He derives the Nymphaeales and Ranunculales separately from the Magnoliales.

A separation of woody from herbaceous groups appears in Endlicher's sequence of families: Annonaceae, Magnoliaceae, Ranunculaceae, Berberidaceae. The sequence of families given by Bentham and Hooker lays greater stress on floral structure. It runs: Ranunculaceae, Magnoliaceae, Annonaceae, Berberidaceae, Nymphaeaceae. Eichler's sequence: Berberidaceae, Annonaceae, Magnoliaceae, Ranunculaceae, Nymphaeaceae brings the Ranunculaceae and Nymphaeaceae closer together. Engler's sequence: Nymphaeaceae, Ranunculaceae, Berberidaceae, Magnoliaceae, Annonaceae, although separating the woody from the herbaceous families, places the herbaceous first. Bessey's sequence: Magnoliaceae, Annonaceae, Ranunculaceae, Berberidaceae follows that of Endlicher in placing the woody families ahead of the herbaceous. It also places the Magnoliaceae ahead of the Annonaceae, and in this respect it follows the scheme of Bentham and Hooker. As for the Nymphaeaceae, Bessey alone disposes them in the

Papaverales. Hutchinson raises most of the families here mentioned to ordinal status, and their disposition has been described above. He includes the Nymphaeaceae within his Ranunculales. Melchior has the sequence: Magnoliaceae, Annonaceae in his Magnoliales and the sequence: Ranunculaceae, Berberidaceae, Nymphaeaceae in his Ranunculales. Cronquist has a similar arrangement, except that his Nymphaeaceae stand in a separate order. Hutchinson, Melchior, and Cronquist agree in splitting the ranalian complex of earlier authors into woody and herbaceous groups; and they further agree in placing the Magnoliaceae ahead of the Annonaceae within the first of these groups, and in placing the Ranunculaceae ahead of the Berberidaceae within the second.

20 4:1

GLOSSARY FOR MAGNOLIALES

Nymphaeales. Order in subclass Agamopetalidae.

  Nelumbia. Classis in cohors Dialypetalae (Endl.)

  Nymphaeales. Order in subclass Magnoliidae (Cronq.)

Ranunculales. Order in subclass Agamopetalidae.

  Ranales. Order in division Herbacidae (Hutch.)

  Ranunculales. Reihe in Unterklasse Archichlamydeae (Melch.)

  Ranunculales. Order in subclass Magnoliidae (Cronq.)

Berberidales. Order in subclass Agamopetalidae.

Berberidales. Order in division Herbaceae (Hutch.)

Magnoliales. Order in subclass Agamopetalidae.

Polycarpicae. Classis in cohors Dialypetalae (Endl.)

Ranales. Cohors in series Thalamiflorae (B.&H.)

In subclassis Polypetalae.

Polycarpicae. Unterreihe in Reihe Aphanocyclicae (Eichl.)

In Klasse Chori- und Apetalae.

Ranales. Series in subclassis Archichlamydeae (Engl.)

Ranales. Order in superorder Apopetalae-Polycarpellatae (Bessey). In subclass Strobiloideae.

Magnoliales. Order in division Lignosae (Hutch.)

Magnoliales. Reihe in Unterklasse Archichlamydeae (Melch.)

Magnoliales. Order in subclass Magnoliidae (Cronq.)

Annonales. Order in subclass Agamopetalidae.

Annonales. Order in division Lignosae (Hutch.)

Nymphaeaceae. Family in order Magnoliales.

Nymphaeaceae. Ordo in classis Nelumbia (Endl.)

Nymphaeaceae. Ordo in cohors Ranales (B.&H.)

Nymphaeaceae. Familie in Unterreihe Polycarpicae (Eichl.)

Nymphaeaceae. Familia in series Ranales (Engl.)

Nymphaeaceae. Family in order Rhoeadales (Bessey).

Nymphaeaceae. Family in order Ranales (Hutch.)

Nymphaeaceae. Familie in Reihe Ranunculales (Melch.)

Nymphaeaceae. Family in order Nymphaeales (Cronq.)

Ranunculaceae. Family in order Magnoliales.

Ranunculaceae. Ordo in classis Polycarpicae (Endl.)

Ranunculaceae. Ordo in cohors Ranales (B.&H.)

Ranunculaceae. Familie in Unterreihe Polycarpicae (Eichl.)

Ranunculaceae. Familia in series Ranales (Engl.)

Ranunculaceae. Family in order Ranales (Bessey), (Hutch.)

Ranunculaceae. Familie in Reihe Ranunculales (Melch.)

Ranunculaceae. Family in order Ranunculales (Cronq.)

Berberidaceae. Family in order Magnoliales.

Berberideae. Ordo in classis Polycarpicae (Endl.)

Berberideae. Ordo in cohors Ranales (B.&H.)

Berberideae. Familie in Unterreihe Polycarpicae (Eichl.)

Berberidaceae. Familia in series Ranales (Engl.)

Berberidaceae. Family in order Ranales (Bessey).

Berberidaceae. Family in order Berberidales (Hutch.)

Berberidaceae. Familie in Reihe Ranunculales (Melch.)

Berberidaceae. Family in order Ranunculales (Cronq.)

Magnoliaceae. Family in order Magnoliales.

Magnoliaceae. Ordo in classis Polycarpicae (Endl.)

Magnoliaceae. Ordo in cohors Ranales (B.&H.)

Magnoliaceae. Familie in Unterreihe Polycarpicae (Eichl.)

Magnoliaceae. Familia in series Ranales (Engl.)

Magnoliaceae. Family in order Ranales (Bessey).

Magnoliaceae. Family in order Magnoliales (Hutch.), (Cronq.)

Magnoliaceae. Familie in Reihe Magnoliales (Melch.)

Annonaceae. Family in order Magnoliales.

Anonaceae. Ordo in classis Polycarpicae (Endl.)

Anonaceae. Ordo in cohors Ranales (B.&H.)

Anonaceae. Familie in Unterreihe Polycarpicae (Eichl.)

Anonaceae. Familia in series Ranales (Engl.)

Anonaceae. Family in order Ranales (Bessey).

Annonaceae. Family in order Annonales (Hutch.)

Annonaceae. Familie in Reihe Magnoliales (Melch.)

Annonaceae. Family in order Magnoliales (Cronq.)

Saururaceae. See 9 4.

Saururaceae. Family in order Ranales (Bessey).

Leitneriaceae. See 12 4.

Leitneriaceae. Family in order Ranales (Bessey).

21 1:1

PAPAVERALES

The Papaverales include the Papaveraceae, Capparaceae, Cruciferae, and Resedaceae. Taxa of the order have the gynoecium composed of two or more carpels united into a compound and unilocular ovary with usually parietal placentation. The most important species of poppy is the introduced Papaver somniferum, from which opium is obtained. This plant has from ancient times been used as a source for various soporifics. Cleome in our area is the best-known genus representing the Capparaceae. The Cruciferae constitute one of the larger plant families, including food plants, ornamentals, and weeds. The Resedaceae are a Mediterranean family sparsely represented in our area. *Reseda odorata*, mignonette, is cultivated as a fragrant ornamental.

Papaver is the ancient Latin name for the poppy, but its etymological signification is obscure. *Capparis* is the ancient Latin name for the caper, and it is further derived from Greek. In Greek the name was applied to the plant now known as *Capparis spinosa*. Endlicher treats *Capparis* as a *d*-stem forming *Capparideae*. Lindley (1836) preferring the termination *-aceae* introduced *Capparidaceae*. Recently this form has been rejected as philologically unsound and *Capparaceae* substituted for it. For an objection to this change see *Taxon* 15: 205 seq. (1966). The family name *Cruciferae* is post-Linnaean, the Linnaean name being *Siliquosae*. *Cruciferae* signifies: "plants bearing a cross," and it is descriptive of the cruciate appearance of the two sets of oppositely paired petals. The name *Reseda* is an ancient Latin name applied to the plant now known as *Reseda alba*. It signifies: "plant from which a medicine is prepared," or more literally: "plant (-*a*) which again (*re*-) settles (-*sed*-) a disturbed condition."

The Papaveraceae commonly include *Fumaria*. However Eichler, Hutchinson, and Cronquist base a separate family on this genus. The name Fumariaceae is not included in the glossary for Papaverales.

21 2:1

PAPAVERACEAE

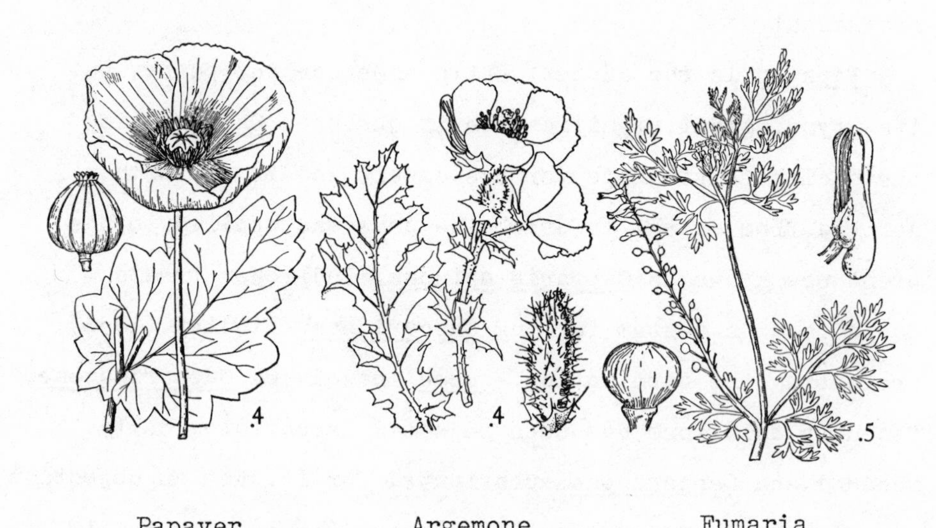

Papaver somniferum	Argemone intermedia	Fumaria officinalis
-	-	-
poppy	prickly poppy	fumitory

PAPAVERACEAE

dHypoG1̸2-nL1AnCa,2̸,4-nK2(-4)/caps.plac.par.juice milky,col.

Dicotyledonous plants with hypogynous flowers. Gynoecium of 2 to many united carpels and a single locule.

Androecium of numerous stamens. Corolla of 4 to many petals, actinomorphic or zygomorphic. Calyx of 2 sepals (or as many as 4). Fruit a capsule. Placentation parietal. Juice milky, or colored.

Papaver sp.

dHypoGi̸8L1AnC4K2/

Fumaria sp.

dHypoGi̸2L1A4(i̸½+1+½)+(i̸½+1+½)Ca̸4(2+2)K2/

Androecium of 4 stamens (in two groups of united parts, each with 1 complete stamen and ½ of each of two divided stamens). Corolla of 4 petals (of two pairs of 2 petals each), zygomorphic.

CAPPARACEAE

dHypoGi̸2L1A6-nCa,a̸,4Ki,_,4/caps.

Dicotyledonous plants with hypogynous flowers. Gynoecium of 2 united carpels and a single locule. Androecium of 6 to many stamens. Corolla of 4 petals, actinomorphic or zygomorphic. Calyx of 4 sepals, free or united at the base. Fruit a capsule.

Cleome sp.

dHypoGi̸2L1A6C4K4/

CAPPARACEAE

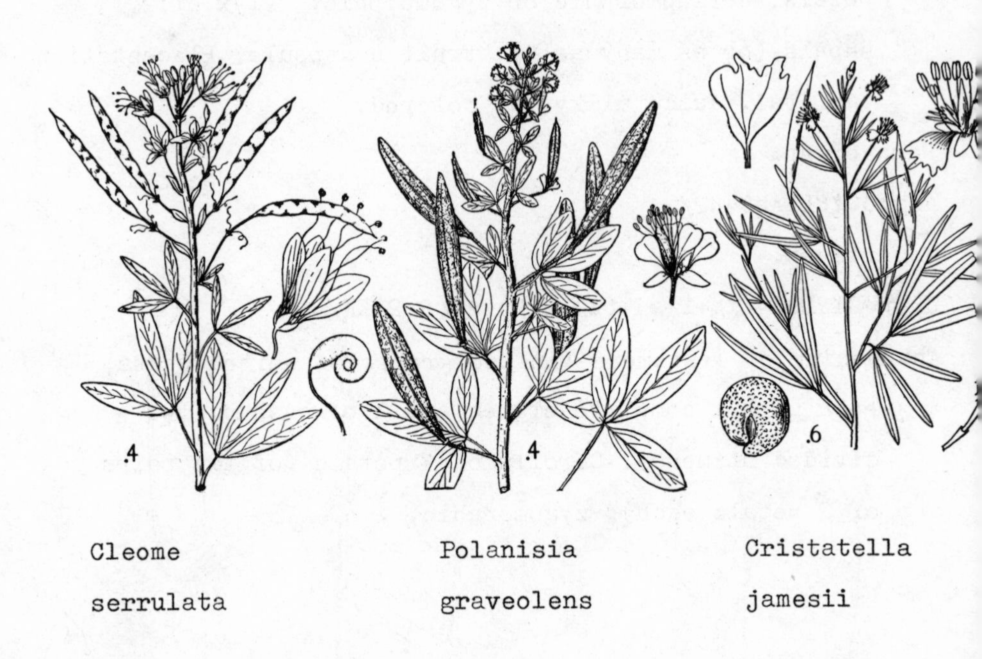

Cleome serrulata	Polanisia graveolens	Cristatella jamesii
-	-	-

CRUCIFERAE

dHypoG/2L2(w.replum)A6(4+2(short))C4K4/silique,silicle, plac.par.

Dicotyledonous plants with hypogynous flowers. Gynoecium of 2 united carpels and 2 locules (separated by a replum). Androecium of 6 stamens (an inner set of 4 and an outer set of 2 (which are shorter than the others)). Corolla of 4 petals. Calyx of 4 sepals.

Fruit a silique, or silicle. Placentation parietal.

Brassica sp.

dHypoGi̸2L2(w.replum)A6(4+2(short))C4K4/

CRUCIFERAE

Brassica nigra	Thlaspi arvense	Capsella bursa-pastoris
black mustard	-	shepherd's
-	penny cress	purse

RESEDACEAE

dHypo(w.gynandroph.disk(poster.enlargem.))Gi̸2-6L1A(3-)n

Cø4-7K4-7/caps.plac.par.

Dicotyledonous plants with hypogynous flowers (having a gynandrophore disk (which has a posterior enlargement). Androecium of numerous stamens (or as few as 3). Corolla of 4 to 7 petals, zygomorphic. Calyx of 4 to 7 sepals. Fruit a capsule. Placentation parietal.

Reseda sp.

dHypo(w.gynandroph.disk(poster.enlargem.))Gi̸3AnCa̸6K6/

RESEDACEAE

Reseda luteola	Reseda lutea	Reseda alba
dyer's rocket	-	-
-	mignonette	mignonette

Endlicher places the Papaverales directly after the Magnoliales toward the beginning of the Polypetalidae, the third of his three dicot subclasses. Bentham and Hooker do not recognize the Papaverales, but they place the papaveraceous family group in the Violales. This order comes directly after the Magnoliales within the Polypetalidae, the first of their three dicot subclasses. Eichler places the Papaverales after the Magnoliales, and the two taxa constitute the first order group in the superorder Aphanocyclicae within the polypetalous portion of the Agamopetalidae, the second of his two dicot subclasses. Engler places the Papaverales after the Magnoliales toward the beginning of the polypetalous portion of the Agamopetalidae, the first of his two dicot subclasses. Bessey places the Papaverales after the Theales toward the end of the polypetalous portion of the Dicotostrobiloididae, the first of his two dicot subclasses. He diagrams as a phylogenetic sequence: Magnoliales, Papaverales. Hutchinson removes some families from the traditional Papaverales and raises them to ordinal rank. He places the residual Papaverales within the first order group of the Herbacidae, the second of his two dicot subclasses. This order group runs: Ranunculales, Berberidales, Papaverales, Cruciferales, Resedales. The last three of these orders may be regarded as constituting a papaveralian order subgroup. Hutchinson diagrams as a phylogenetic sequence: Papaverales, Cruciferales, Resedales, and he derives the sequence directly from the Ranunculales. Melchior places the Papaverales within the polypetalous portion of the Agamopetalidae, the first of his two dicot subclasses. Cronquist places the Papaverales at the end of the Magnoliidae, the first of his six dicot subclasses. He diagrams as a phylogenetic sequence: Magnoliales, Ranunculales, Papaverales. He also recognizes an order Capparales which he places within the Theidae, the fourth of these subclasses.

Endlicher's Papaverales have the sequence: Papaveraceae, Cruciferae, Capparaceae, Resedaceae. This is also the Bentham and Hooker sequence. Eichler too adopts it, but he gives separate family status to the Fumariaceae directly after the Papaveraceae. Engler and Prantl adopt the Endlicher sequence, but Engler in later versions of his system transposes the positions of the Cruciferae and Capparaceae. Bessey alone introduces the Nymphaeaceae into the order and specifies the sequence: Papaveraceae, Nymphaeaceae, Resedaceae, Capparaceae, Cruciferae. Hutchinson's Papaverales include the Papaveraceae and Fumariaceae. His Cruciferales include the Cruciferae and Cleomaceae. His Resedales include the Resedaceae. His Capparaceae is a family of his subclass Lignosidae and is not represented in our area. Melchior follows the later

Englerian sequence of papaveralian families. Cronquist's Papaverales include the Papaveraceae and Fumariaceae. His Capparales have the sequence: Capparaceae, Cruciferae, Resedaceae.

21 4:1

GLOSSARY FOR PAPAVERALES

Papaverales. Order in subclass Agamopetalidae.

Rhoeadeae. Classis in cohors Dialypetalae (Endl.)

Rhoeadinae. Unterreihe in Reihe Aphanocyclicae (Eichl.) In Klasse Chori- und Apetalae.

Rhoeadales. Series in subclassis Archichlamydeae (Engl.)

Rhoeadales. Order in superorder Apopetalae-Polycarpellatae (Bessey). In subclass Strobiloideae.

Rhoeadales. Order in division Herbaceae (Hutch.)

Papaverales. Reihe in Unterklasse Archichlamydeae (Melch.)

Papaverales. Order in subclass Magnoliidae (Cronq.)

Capparales (Cruciferales). Order in subclass Agamopetalidae.

Cruciales. Order in division Herbaceae (Hutch.)

(Capparidales. Order in division Lignosae (Hutch.) Not represented in our area.)

Capparales. Order in subclass Dilleniidae (Cronq.)

Resedales. Order in subclass Agamopetalidae.

Resedales. Order in division Herbaceae (Hutch.)

Papaveraceae. Family in order Papaverales.

Papaveraceae. Ordo in classis Rhoeadeae (Endl.)

Papaveraceae. Ordo in cohors Parietales (B.&H.)

Papaveraceae. Familie in Unterreihe Rhoeadinae (Eichl.)

Papaveraceae. Familia in series Rhoeadales (Engl.)

Papaveraceae. Family in order Rhoeadales (Bessey), (Hutch.)

Papaveraceae. Familie in Reihe Papaverales (Melch.)

Papaveraceae. Family in order Papaverales (Cronq.)

Capparaceae (Cleomaceae). Family in order Papaverales.

Capparideae. Ordo in classis Rhoeadeae (Endl.)

Capparideae. Ordo in cohors Parietales (B.&H.)

Capparideae. Familie in Unterreihe Rhoeadinae (Eichl.)

Capparidaceae. Familia in series Rhoeadales (Engl.)

Capparidaceae. Family in order Rhoeadales (Bessey).

Cleomaceae. Family in order Rhoeadales (Hutch.)

(Capparidaceae. Family in order Capparidales (Hutch.) Not represented in our area.)

Capparaceae. Familie in Reihe Papaverales (Melch.)

Capparaceae. Family in order Capparales (Cronq.)

Cruciferae. Family in order Papaverales.

Cruciferae. Ordo in classis Rhoeadeae (Endl.)

Cruciferae. Ordo in cohors Parietales (B.&H.)

Cruciferae. Familie in Unterreihe Rhoeadinae (Eichl.)

Cruciferae. Familia in series Rhoeadales (Engl.)

Brassicaceae. Family in order Rhoeadales (Bessey).

Brassicaceae. Family in order Cruciales (Hutch.)

Cruciferae. Familie in Reihe Papaverales (Melch.)

Cruciferae. Family in order Capparales (Cronq.)

Resedaceae. Family in order Papaverales.

Resedaceae. Ordo in classis Rhoeadeae (Endl.)

Resedaceae. Ordo in cohors Parietales (B.&H.)

Resedaceae. Familie in Unterreihe Rhoeadinae (Eichl.)

Resedaceae. Familia in series Rhoeadales (Engl.)

Resedaceae. Family in order Rhoeadales (Bessey).

Resedaceae. Family in order Resedales (Hutch.)

Resedaceae. Familie in Reihe Papaverales (Melch.)

Resedaceae. Family in order Capparales (Cronq.)

Nymphaeaceae. See 20 4.

Nymphaeaceae. Family in order Rhoeadales (Bessey).

22 1:1

SARRACENIALES

The order Sarraceniales is represented in our area by <u>Sarracenia</u> in the family Sarraceniaceae and by <u>Drosera</u> in the family Droseraceae. The Sarraceniales are insectivorous plants more easily identified by their highly modified leaves than by their flowers. The group is not prominent in the flora, and it is not given regular coverage in the present survey. The genus Sarracenia is characterized by pitcher-like leaves. The genus Drosera is characterized by leaves having mucilaginous hairs. Plants of this group are cultivated as curiosities.

Tournefort based the name Sarracenia on Sarracin, a Latinized form of Sarrazin, the name of a Quebec physician who had sent him specimens and a description of one of the species of this genus. Drosera is a substantivized form of the nominative singular feminine of the ancient Greek adjective signifying "dewy." The adjective droseros is derived in turn from the noun drosos, "dew."

22 2:1

SARRACENIACEAE

dHypoGi̸5(w.umbrella-shaped style)AnC5,O,K,P,5/perf., unisex.,caps.carniv.

Dicotyledonous plants with hypogynous flowers. Gynoecium of 5 united carpels (with an umbrella-shaped style). Androecium of numerous stamens. Corolla of 5 petals, or none. Calyx, or perianth, of 5 sepals, or tepals. Flowers perfect, or unisexual. Fruit a capsule. Plants carnivorous.

Sarracenia sp.

dHypoGi̸5(w.umbrella-shaped style)AnC5K5/perf.

SARRACENIACEAE

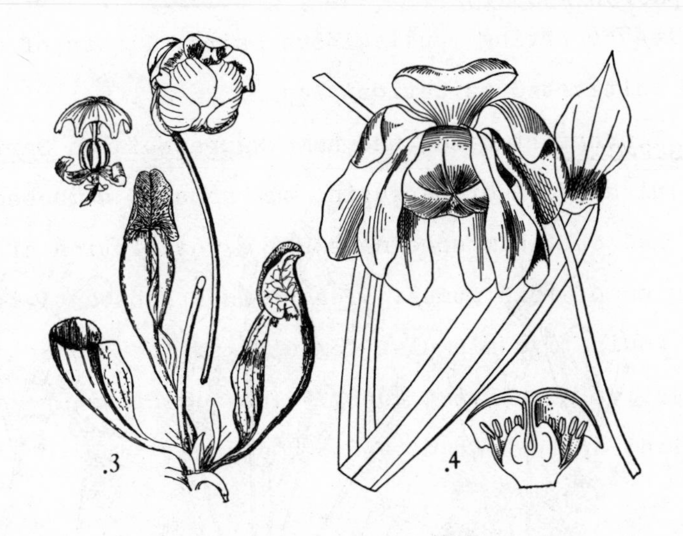

Sarracenia purpurea	Sarracenia flava
pitcher plant	trumpet leaf
-	-

DROSERACEAE

dHypoG/3-5L1A5-nC5K/5/caps./plac.par./carniv.

Dicotyledonous plants with hypogynous flowers. Gynoecium of 3 to 5 united carpels and a single locule. Androecium of 5 to many stamens. Corolla of 5 petals. Calyx of 5 united sepals. Fruit a capsule. Placentation parietal. Plants carnivorous.

Drosera sp.

dHypoGi̸3L1A5C5Ki̸5/

DROSERACEAE

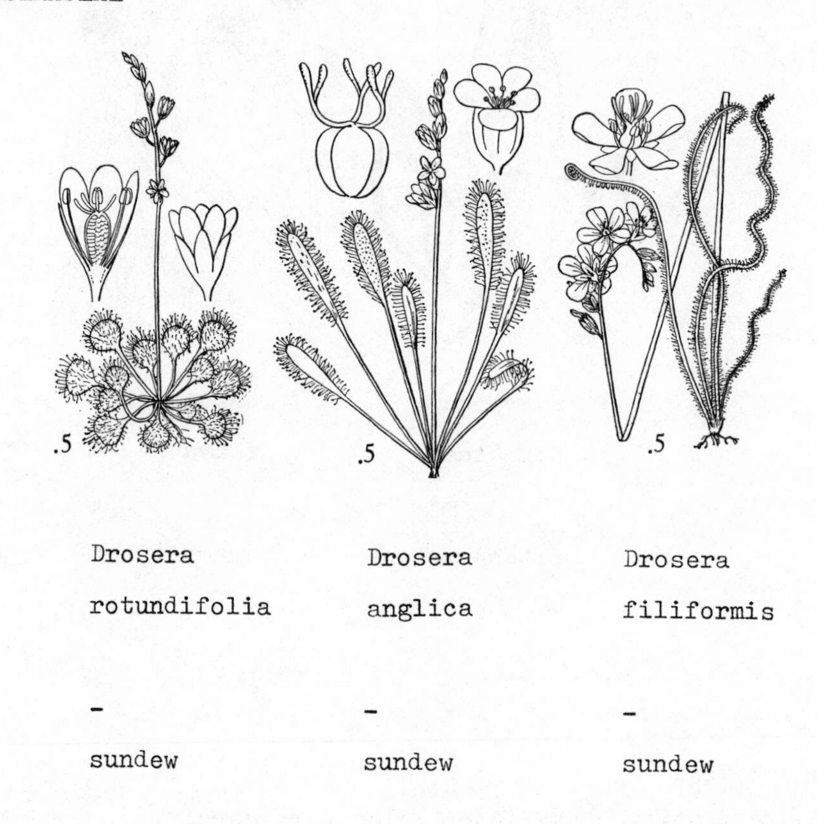

Drosera rotundifolia	Drosera anglica	Drosera filiformis
-	-	-
sundew	sundew	sundew

22 3:1

Endlicher does not recognize the Sarraceniales but classifies <u>Sarracenia</u> in the Nymphaeaceae within the Nymphaeales. He places the Droseraceae in the Violales. Bentham and Hooker also do not recognize the Sarraceniales, but classify the Sarraceniaceae within the Violales. The Droseraceae, however, they classify in the Rosales remote from the Violales. Eichler similarly does not recognize the Sarraceniales, but he classifies both the Sarraceniaceae and Droseraceae within his Theales, a group inclusive of the Violaceae. Engler establishes the Sarraceniales between the Papaverales and Rosales. Bessey also accepts the Sarraceniales, but he places them after the Malvales. He diagrams as a phylogenetic sequence: Magnoliales, Malvales, Sarraceniales. Hutchinson places the Sarraceniales after the Saxifragales, and he diagrams as a phylogenetic sequence: Saxifragales, Sarraceniales. Melchior places the Sarraceniales after the Theales. Cronquist places the Sarraceniales within his subclass Theidae, and he diagrams as a phylogenetic sequence: Theales, Sarraceniales.

22 4:1

GLOSSARY FOR SARRACENIALES

Sarraceniales. Order in subclass Agamopetalidae.

Sarraceniales. Series in subclassis Archichlamydeae (Engl.)

Sarraceniales. Order in superorder Apopetalae-Polycarpellatae (Bessey). In subclass Strobiloideae.

Sarraceniales. Order in division Herbaceae (Hutch.)

Sarraceniales. Reihe in Unterklasse Archychlamydeae (Melch.)

Sarraceniales. Order in subclass Dilleniidae (Cronq.)

Sarraceniaceae. Family in order Sarraceniales.

Sarraceniaceae. Ordo in cohors Parietales (B.&H.)

Sarraceniaceae. Familie in Unterreihe Cistiflorae (Eichl.)

Sarraceniaceae. Familia in series Sarraceniales (Engl.)

Sarraceniaceae. Family in order Sarraceniales (Bessey), (Hutch.), (Cronq.)

Sarraceniaceae. Familie in Reihe Sarraceniales (Melch.)

Droseraceae. Family in order Sarraceniales.

Droseraceae. Ordo in classis Parietales (Endl.)

Droseraceae. Ordo in cohors Rosales (B.&H.)

Droseraceae. Familie in Unterreihe Cistiflorae (Eichl.)

Droseraceae. Familia in series Sarraceniales (Engl.)

Droseraceae. Family in order Rosales (Bessey).

Droseraceae. Family in order Sarraceniales (Hutch.), (Cronq.)

Droseraceae. Familie in Reihe Sarraceniales (Melch.)

PODOSTEMALES

The order Podostemales is represented in our area only by the species Podostemum ceratophyllum. The group is not given regular coverage in the present survey. Reduction in the order has been so extreme that its affinities are obscure. The foliage has seaweed characteristics, and the flowers are apetalous. Systems differ as to whether this apetalous condition be primary or secondary. The generic name Podostemum was used by Michaux in 1803. It signifies "plant with the stamens on a single footing," in reference to its unilateral androecium. The family name Podostemaceae and the ordinal name Podostemales are based on this generic name. An alternative form Podostemon appears based on a Greek root for "thread" rather than the one for "stamen," and on it the names Podostemonaceae and Podostemonales were based, but these are at present relegated to botanical synonymy.

23 2:1

PODOSTEMACEAE

dHypoG/2-3A2-nP/2-3,0/caps.thalloid

Dicotyledonous plants with hypogynous flowers. Gynoecium of 2 to 3 united carpels. Androecium of 2 to many

PODOSTEMACEAE

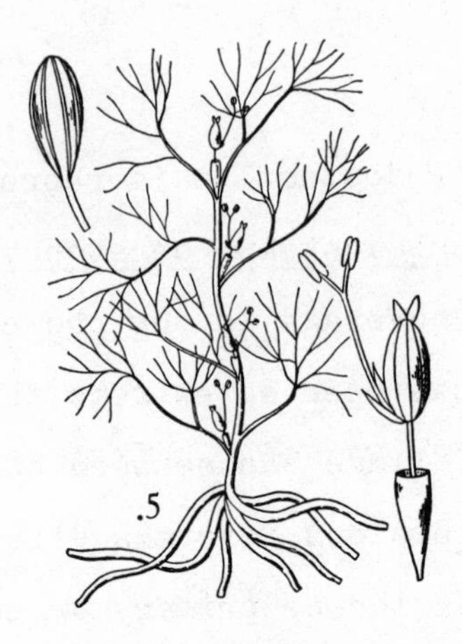

Podostemum
ceratophyllum

-

riverweed

stamens. Perianth of 2 to 3 united tepals, or none. Fruit a capsule. Plants thalloid.

Podostemum ceratophyllum

dHypo(w.gynandroph.)G/2A2(on unilat.gynandroph.branch)S2/

Dicotyledonous plants with hypogynous flowers (having a gynandrophore). Androecium of 2 stamens (on a unilateral gynandrophore branch). Staminodia 2.

Endlicher, Eichler, Engler, and Bessey do not recognize the Podostemales, but they differ in their placement of the Podostemaceae. Endlicher classifies the Podostemaceae in the Urticales. Bentham and Hooker classify the Podostemales in the subclass Apetalidae. In the Blüthendiagramme Eichler does not discuss the Podostemaceae, but elsewhere he associates the group with the Saxifragales. Engler adopts a similar view by incorporating the Podostemaceae within the Rosales. The Engler and Diels Syllabus of 1936, however, reverts to Endlicher's views by establishing the Podostemales after the Urticales. Fernald, although committed to the Engler and Prantl (Dalla Torre and Harms) sequence of families, manages to extricate the Podostemaceae from the Rosales by recognizing the Podostemales as a separate order. Bessey places the Podostemaceae within the Caryophyllales. Hutchinson places the Podostemales after the Sarraceniales within the subclass Herbacidae, and by diagram he confirms this as a phylogenetic sequence. Melchior places the Podostemales after the Rosales. Cronquist places the Podostemales in the subclass Rosidae, and he diagrams as a phylogenetic sequence: Rosales, Podostemales.

GLOSSARY FOR PODOSTEMALES

Podostemales. Order in subclass Agamopetalidae.

Aquaticae. Classis in cohors Apetalae (Endl.)

Multiovulatae aquaticae. Series in subclassis Monochlamydeae (B.&H.)

Podostemales. Order in division Herbaceae (Hutch.)

Podostemales. Reihe in Unterklasse Archichlamydeae (Melch.)

Podostemales. Order in subclass Rosidae (Cronq.)

Podostemaceae. Family in order Podostemales.

Podostemmeae. Ordo in classis Aquaticae (Endl.)

Podostemaceae. Ordo in series Multiovulatae aquaticae (B.&H.)

Podostemaceae. Familie in Unterreihe Saxifraginae (Eichl.)

Podostemonaceae. Familia in series Rosales (Engl.)

Podostemonaceae. Family in order Caryophyllales (Bessey).

Podostemaceae. Family in order Podostemales (Hutch.), (Cronq.)

Podostemaceae. Familie in Reihe Podostemales (Melch.)

ROSALES

The order Rosales contains the families Crassulaceae, Saxifragaceae, Hamamelidaceae, Platanaceae, Rosaceae, and Leguminosae. The Crassulaceae are succulents with glands at the base of their usually distinct carpels. The carpels are isomeric with the sepals. Endosperm is present. The Saxifragaceae in the broader circumscription are a diverse aggregation characterized by flowers with united carpels fewer than the sepals, each carpel having two ovules. Endosperm present. The Platanaceae are monoecious trees with flowers in capitate inflorescences. The gynoecia have several distinct carpels. Endosperm none. The Rosaceae have perfect flowers with commonly numerous carpels. Endosperm none. The Leguminosae have mostly irregular perfect flowers with a single carpel. Endosperm none.

All of the above family names except *Leguminosae* are derivative from generic names. *Crassula* is a medieval Latin diminutive feminine noun derived from the Latin adjective *crassus*, "thick." It signifies: "small herb with thick leaves," and was first applied to species of *Sedum*. The adjective *saxifragus*, signifying in ancient Latin "stone breaking or dissolving," was applied to the medicinal use of *Adiantum* in order to dissolve stones in the bladder. The

English word saxifragous, "dissolving bladder stones," is now obsolete, being replaced by the equivalent word with Greek roots: lithotriptic. Saxifraga in modern botanical usage is usually apprehended to signify etymologically simply: "plant which cuts stone," in reference to occurence in cracks of stone. The English name stonecrop occurs in early usage with reference to Crassula. Hamamelis is a Greek name anciently applied to a plant now classified in the Rosaceae. Platanus is the ancient Latin name derived from Greek for the tree now named Platanus orientalis. The name signifies: "plant that has breadth," but whether this refers to crown, or leaves, or pieces of shed bark is disputed. The English word plane occurring in plane tree is derived from the Latin platanus. Rosa is an ancient Latin name derived from Greek and of doubtful etymological signification. Leguminosae is a descriptive family name derived from Latin legumen "pulse," or more literally "fruit gathered by picking," especially pod fruits.

Hydrangea, Philadelphus, and Grossularia are commonly included in the Saxifragaceae. Endlicher recognizes a Grossulariaceae (his Ribesiaceae); and he also recognizes a Philadelphaceae (his Philadelpheae), which last he classifies remote from the Saxifragaceae. Bessey and Cronquist recognize the Hydrangeaceae (incl. Philadelphus) and the Grossulariaceae. Hutchinson not only recognizes the Hydrangeaceae, Philadelphaceae, and Grossulariaceae;

but he establishes them in the order Hydrangeales. Names applicable to these three families are not included in the glossary for Rosales.

The pomaceous and drupaceous rosacean taxa are commonly included in the Rosaceae. Endlicher and Bessey recognize pomaceous and drupaceous families (Endlicher's Pomaceae and Amygdaleae and Bessey's Malaceae and Prunaceae). Names applicable to these two families are not included in the glossary for Rosales.

24 2:1

CRASSULACEAE

Sedum ternatum	Sempervivum tectorum	Tillaea aquatica
-	-	-
stonecrop	houseleek	pigmyweed

CRASSULACEAE

dHypo,Peri,G4-5(usu.scale behind each)A4-5,8-10(oft. obdipl.),Ci,_,4-5,0,K_4-5/fol./succ.

Dicotyledonous plants with hypogynous, or perigynous, flowers. Gynoecium of 4 to 5 carpels (usually a scale behind each). Androecium of 4 to 5, or of 8 to 10 (often obdiplostemonous) stamens. Corolla of 4 to 5 petals, free or united at the base, or none. Calyx of 4 to 5 tepals united at the base. Fruit a follicle. Plants succulent.

Sedum sp.

dHypoG5(scale behind each)A10C5K_5/

SAXIFRAGACEAE

dPeri,Epi,(Hypo),Gi̸2(-5)L2(-5),1A4-5,8-10,(n),C4-5,0,Ki̸,_,(i),4-5/caps.,ber./styles i,±i̸,ovules n per L

Dicotyledonous plants with perigynous, or epigynous (or less frequently hypogynous) flowers. Gynoecium of 2 united carpels (or as many as 5) and the locules isomeric with the carpels, or only 1. Androecium of 4 to 5, or 8 to 10 (or numerous) stamens. Corolla of 4 to 5 petals, or none. Calyx of 4 to 5 sepals, united or coherent at the base (or less frequently free). Fruit a capsule, or a berry. Styles free, or more or less united. Ovules many per locule.

SAXIFRAGACEAE

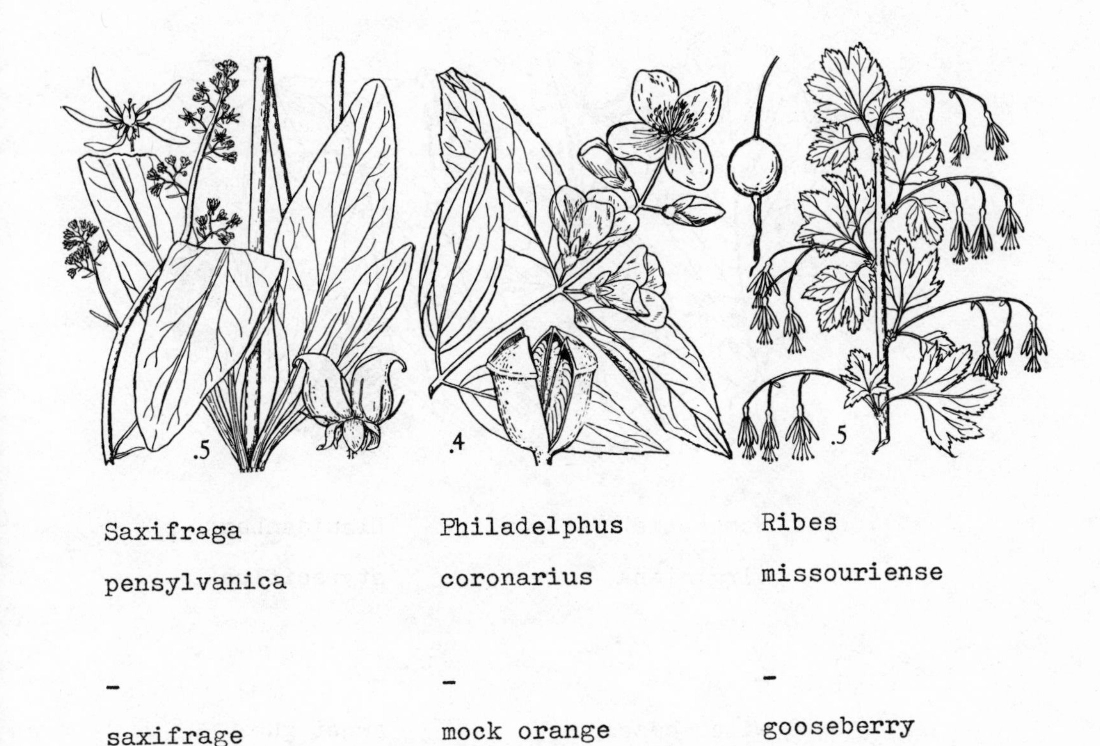

Saxifraga pensylvanica	Philadelphus coronarius	Ribes missouriense
-	-	-
saxifrage	mock orange	gooseberry

Saxifraga sp.

dPeriG/2A10(obdipl.)C5K5/

Ribes sp.

dPeriG/2L1A5C5K5/

HAMAMELIDACEAE

dPeri,Psilo,G/2A4-nC4-5,0,K/4-5,0/perf.,monoec.,woody caps.

styles i,±/

HAMAMELIDACEAE

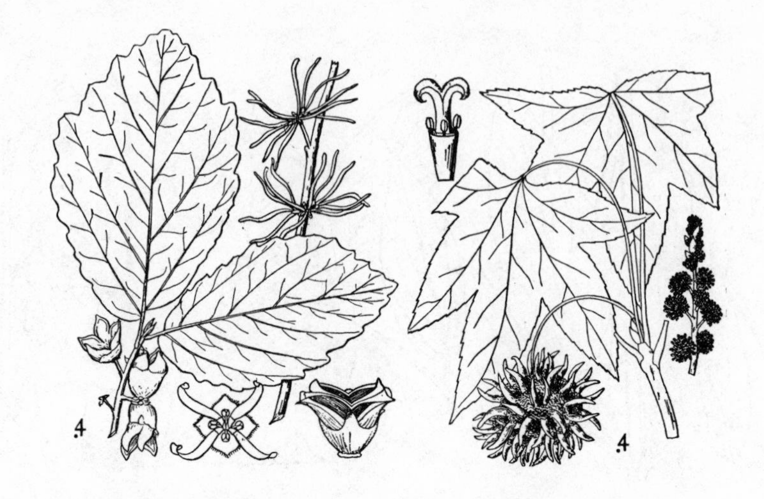

Hamamelis virginiana	Liquidambar styraciflua
-	-
witch hazel	sweet gum

Dicotyledonous plants with perigynous, or psilogynous, flowers. Gynoecium of 2 united carpels. Androecium of 4 to many stamens. Corolla of 4 to 5 petals, or none. Calyx of 4 to 5 united sepals, or none. Flowers perfect, or unisexual and the plants monoecious. Fruit a woody capsule. Styles free, or more or less united.

Hamamelis sp.

dPeriGi̸2A4S4C4Ki̸4/

PLATANACEAE

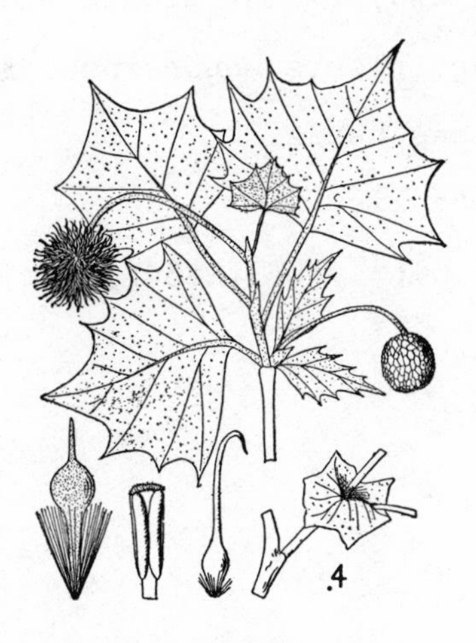

Platanus
occidentalis

-

sycamore

PLATANACEAE

dHypo,Peri,G5-9A3-4(-7),S3-4(f),C3-4(-7)(m),O(f),K/3-5
(-7(m))/monoec.nutlet/heads

Dicotyledonous plants with hypogynous, or perigynous, flowers. Gynoecium of 5 to 9 carpels. Androecium of 3 to 4 stamens (or as many as 7), or staminodia 3 to 4 (in pistillate flowers). Corolla of 3 to 4 petals (or as many as 7) (in staminate flowers), or none (in pistillate flowers). Calyx of 3 to 5 united tepals (or as many as 7 (in staminate flowers)). Flowers unisexual and plants monoecious. Fruit a nutlet. Inflorescences heads.

Platanus occidentalis

dHypoG6A4,S3(f),C4(m),O(f),K/3(f),4(m)/monoec.

ROSACEAE

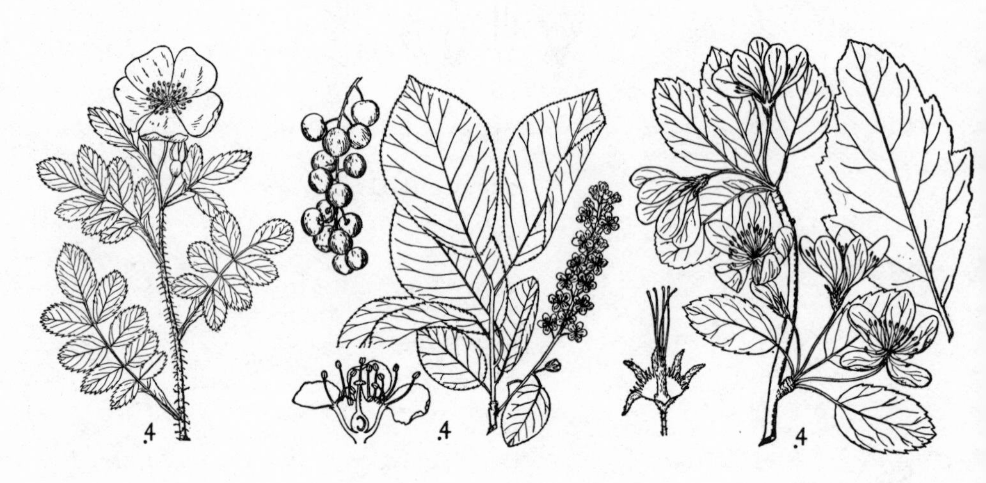

Rosa arkansana var. suffulta	Prunus virginiana	Malus ioensis
-	choke cherry	crab
rose	-	apple

ROSACEAE

dPeri,Epi,Gi,i̸,1-nA4-nC(0-)5(-10)K(4-)5(-10)/ach.,drup., fol.,pome

Dicotyledonous plants with perigynous, or epigynous, flowers. Gynoecium of 1 to many carpels, free or united. Androecium of 4 to many stamens. Corolla of 5 petals (or fewer and even none, or as many as 10). Calyx of 5 sepals (or as few as 4, or as many as 10). Fruit an achene, or drupaceous, or a follicle, or a pome.

Rosa sp.

dPeriGnAnC5K5/

Prunus sp.

dPeriG1AnC5K5/

Malus sp.

dEpiG5AnC5K5/

LEGUMINOSAE

dHypo,Peri,G1Ai̸,i,3-10,n,∠i̸,i,10;i̸9±i17Ci̸,i,a̸,a,4-5 ∠i̸,i,5;i̸2+i37Ki̸,i,4-5/legume(,loment,indeh.)

Dicotyledonous plants with hypogynous, or perigynous, flowers. Gynoecium of a single carpel. Androecium of 3 to 10, or many, stamens, united or free. ∠Commonly 10, united or free; or 9 united and 1 more or less free.7 Corolla of 4 to 5 petals, united

LEGUMINOSAE

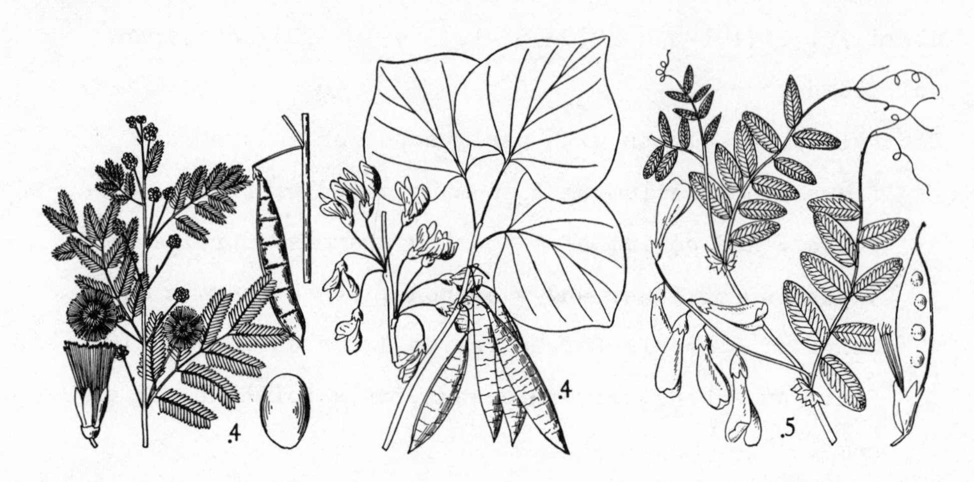

Acacia	Cercis	Vicia
angustissima	canadensis	americana
-	-	-
	redbud	vetch

or free, zygomorphic or actinomorphic. /Commonly 5 united or free; or 2 united and 3 free.7 Calyx of 4 to 5 petals, united or free. Fruit a legume (,or a loment, or indehiscent).

Acacia sp.

dHypoGiAnC/4K/4/

Cercis sp.

dPeriGlAlOC/5(2+1+2)K/5/

Corolla zygomorphic of 5 petals (in 2 and 1 and 2
arrangement (the 2 lateral petals outside the upper
one, and these 3 above the other 2)).

Vicia sp.

dPeriG1A10(i̸9+±i1)Cz̸5(1+2+i̸2)Ki̸5/

Androecium of 10 stamens (9 united and 1 more or less free).
Corolla zygomorphic, of 5 petals (in 1 and 2 and 2
(united) arrangement (the upper petal outside the 2
lateral ones, and these 3 above the other 2, which
are united)).

24 3:1

Comparison of the various systems in their treatment of the rosalian taxa is complex. We must consider several aspects of classification: (1) subclasses, (2) orders, (3) families, (4) sequences of families.

Endlicher, Bentham and Hooker, Eichler, Engler, and Melchior classify the Rosales in either the subclass Polypetalidae or in the polypetalous portion of the subclass Agamopetalidae. Bessey classifies the Rosales as the first and basic order of his subclass Dicotocotyloididae. Hutchinson classifies the Rosales in his subclass Lignosidae. Cronquist classifies the Rosales in his subclass Rosidae. Some of these systems recognize additional orders for rosalian taxa, but most of these fall in the above mentioned subclasses. However Hutchinson classifies the Saxifragales in his subclass Herbacidae, and Cronquist classifies the Hamamelidales in his subclass Hamamelidae.

Bentham and Hooker, Engler, Bessey, and Melchior give the Rosales broad circumscription. Endlicher, Eichler, and Hutchinson recognize also the Saxifragales and Legumenales. Endlicher places the Hamamelidaceae in the Umbellales, but Hutchinson and Cronquist recognize an order Hamamelidales inclusive of the Hamamelidaceae and Platanaceae. In other systems the Hamamelidaceae are otherwise accounted for. Eichler assigns them to the Saxifragales, but Bentham and Hooker, Engler, Bessey, and Melchior assign them to the Rosales. As for the Platanaceae, Endlicher, Bentham and Hooker, and Eichler place them in the Urticales, but Engler, Bessey, and Melchior put them in the Rosales.

The Saxifragaceae, Rosaceae, and Leguminosae are variously delimited. Bentham and Hooker, Eichler, Engler, and Melchior give the Saxifragaceae broad circumscription. Endlicher, Bessey, Hutchinson, and Cronquist, however, separate the Grossulariaceae from the Saxifragaceae, and they deal variously with the Hydrangeaceae and Philadelphaceae. See further under 24 1. Most of the systems give the Rosaceae broad circumscription. Endlicher and Bessey, however, split off the Pomaceae and Drupaceae. See further under 24 1. Bentham and Hooker, Engler, Melchior, and Cronquist give the Leguminosae broad circumscription. Endlicher, Eichler, Bessey, and Hutchinson, however, subdivide this group into smaller families. Endlicher recognizes the Mimosaceae and Papilionaceae. Eichler, Bessey, and Hutchinson accept both, and they split off the Caesalpiniaceae from Endlicher's Papilionaceae.

The systems with broad circumscription of the Rosales differ in their sequence of families. That of Bentham and Hooker puts the Leguminosae in first position so that the Rosaceae in second position stand adjacent to both the Leguminosae and the Saxifragaceae in third position. Further on come the Droseraceae and Hamamelidaceae. Engler reverses the sequence for the first three of these families. This gives: Saxifragaceae, Rosaceae, Leguminosae as a sequence. He also inserts the Hamamelidaceae after the Saxifragaceae. Bessey starts his sequence of rosalian families with a rosacean group, followed by a leguminosean group, followed in turn by a saxifragacean group. Further on come the Droseraceae and Hamamelidaceae. Melchior adopts the Englerian sequence: Saxifragaceae, Rosaceae, Leguminosae, but he positions the Hamamelidaceae before these. Cronquist also accepts the Englerian Saxifragaceae, Rosaceae, Leguminosae sequence, but he classifies the Hamamelidaceae elsewhere.

In all of the systems the Crassulaceae are classified close to the Saxifragaceae. Endlicher, Eichler, and Hutchinson recognize the Saxifragales as containing the Crassulaceae as well as the Saxifragaceae.

GLOSSARY FOR ROSALES

Saxifragales. Order in subclass Agamopetalidae.

Corniculatae. Classis in cohors Dialypetalae (Endl.)

Saxifraginae. Unterreihe in Reihe Calyciflorae (Eichl.)

In Klasse Chori- und Apetalae.

Saxifragales. Order in division Herbaceae (Hutch.)

Hamamelidales. Order in subclass Agamopetalidae.

Hamamelidales. Order in division Lignosae (Hutch.)

Hamamelidales. Order in subclass Hamamelidae (Cronq.)

Rosales. Order in subclass Agamopetalidae.

Rosiflorae. Classis in cohors Dialypetalae (Endl.)

Rosales. Cohors in series Calyciflorae (B.&H.)

In subclassis Polypetalae.

Rosiflorae. Unterreihe in Reihe Calyciflorae (Eichl.)

In Klasse Chori- und Apetalae.

Rosales. Series in subclassis Archichlamydeae (Engl.)

Rosales. Order in superorder Apopetalae (Bessey).

In subclass Cotyloideae.

Rosales. Order in division Lignosae (Hutch.)

Rosales. Reihe in Unterklasse Archichlamydeae (Melch.)

Rosales. Order in subclass Rosidae (Cronq.)

Leguminales. Order in subclass Agamopetalidae.

Leguminosae. Classis in cohors Dialypetalae (Endl.)

Leguminosae. Unterreihe in Reihe Calyciflorae (Eichl.)
In Klasse Chori- und Apetalae.

Leguminales. Order in division Lignosae (Hutch.)

Crassulaceae. Family in order Rosales.

Crassulaceae. Ordo in classis Corniculatae (Endl.)

Crassulaceae. Ordo in cohors Rosales (B.&H.)

Crassulaceae. Familie in Unterreihe Saxifraginae (Eichl.)

Crassulaceae. Familia in series Rosales (Engl.)

Crassulaceae. Family in order Rosales (Bessey), (Cronq.)

Crassulaceae. Family in order Saxifragales (Hutch.)

Crassulaceae. Familie in Reihe Rosales (Melch.)

Saxifragaceae. Family in order Rosales.

Saxifragaceae. Ordo in classis Corniculatae (Endl.)

Saxifrageae. Ordo in cohors Rosales (B.&H.)

Saxifragaceae. Familie in Unterreihe Saxifraginae (Eichl.)

Saxifragaceae. Familia in series Rosales (Engl.)

Saxifragaceae. Family in order Rosales (Bessey), (Cronq.)

Saxifragaceae. Family in order Saxifragales (Hutch.)

Saxifragaceae. Familie in Reihe Rosales (Melch.)

Hamamelidaceae. Family in order Rosales.

Hamamelideae. Ordo in classis Discanthae (Endl.)

Hamamelideae. Ordo in cohors Rosales (B.&H.)

Hamamelideae. Familie in Unterreihe Saxifraginae (Eichl.)

Hamamelidaceae. Familia in series Rosales (Engl.)

Hamamelidaceae. Family in order Rosales (Bessey).

Hamamelidaceae. Family in order Hamamelidales (Hutch.),
(Cronq.)

Hamamelidaceae. Familie in Reihe Rosales (Melch.)

Platanaceae. Family in order Rosales.

Plataneae. Ordo in classis Juliflorae (Endl.)

Platanaceae. Ordo in series Unisexuales (B.&H.)

Platanaceae. Familie in Unterreihe Urticinae (Eichl.)

Platanaceae. Familia in series Rosales (Engl.)

Platanaceae. Family in order Rosales (Bessey).

Platanaceae. Family in order Hamamelidales (Hutch.), (Cronq.)

Platanaceae. Familie in Reihe Rosales (Melch.)

Rosaceae. Family in order Rosales.

Rosaceae. Ordo in classis Rosiflorae (Endl.)

Rosaceae. Ordo in cohors Rosales (B.&H.)

Rosaceae. Familie in Unterreihe Rosiflorae (Eichl.)

Rosaceae. Familia in series Rosales (Engl.)

Rosaceae. Family in order Rosales (Bessey), (Hutch.), (Cronq.)

Rosaceae. Familie in Reihe Rosales (Melch.)

Leguminosae. Family in order Rosales.

Papilionaceae (incl. tribus Caesalpineae), Mimoseae. Ordines in classis Leguminosae (Endl.)

Leguminosae. Ordo in cohors Rosales (B.&H.)

Papilionaceae, Caesalpiniaceae, Mimosaceae. Familien in Unterreihe Leguminosae (Eichl.)

Leguminosae. Familia in series Rosales (Engl.)

Mimosaceae, Cassiaceae, Fabaceae. Families in order Rosales (Bessey).

Caesalpiniaceae, Mimosaceae, Fabaceae. Families in order Leguminales (Hutch.)

Leguminosae. Familie in Reihe Rosales (Melch.)

Leguminosae. Family in order Rosales (Cronq.)

Droseraceae. See 22 4.

Droseraceae. Ordo in cohors Rosales (B.&H.)

Droseraceae. Family in order Rosales (Bessey).

25 1:1

GERANIALES

The Geraniales include the families Linaceae, Oxalidaceae, Geraniaceae, Rutaceae, and Euphorbiaceae. The Linaceae have a monadelphous androecium of five stamens alternating with staminodia. The Oxalidaceae have trifoliate leaves and a monadelphous androecium with five long stamens alternating with shorter stamens, and the latter alternate with the petals. The Geraniaceae have a variable number of stamens, but the fruiting gynoecium is distinctive in that the carpels which surround an extended beak-like floral axis split away from this axis, the styles coiling and remaining attached to the axis apex. The Rutaceae have glandular-dotted leaves and a distinctive hesperidium berry fruit. The Euphorbiaceae have unisexual flowers aggregated into groups known as cyathia, which superficially resemble perfect flowers.

Linum is the ancient Latin name for flax and the plant from which flax is derived. Oxalis is derived from an ancient Greek word for "sour wine." The word was anciently applied to Rumex acetosa, a salad plant. The use of the word Oxalis to designate sorrel is pre-Linnaean, and it alludes to the sour juice of the plants. The root of the English name sorrel also signifies "sour" and is related to but not derived from the English word sour. Geranium is derived from the ancient Greek name signifying: "plant resembling a crane," in allusion to the pointed aspect of the plant fruit and the crane bill. The element geran- and the name crane are philologically cognate. Note corresponding letters as follows: g/c, r/r, a/a, n/n. Ruta is an ancient Latin name for the plant now named Ruta graveolens, once important medicinally. Euphorbia is a name replacing ancient Latin euphorbea and euphorbeum, words derived from ancient Greek. The Greek name literally signifies "well nourished," but ancient sources attribute to the plant name the signification: "plant containing ingredients of a medicine prepared by the physician Euphorbus." The plant in question is now known as Euphorbia resenifera. The medicinal function of the preparation is expressed in the English name *spurge*, derived from Latin *expurgāre* "purge."

25 2:1

LINACEAE

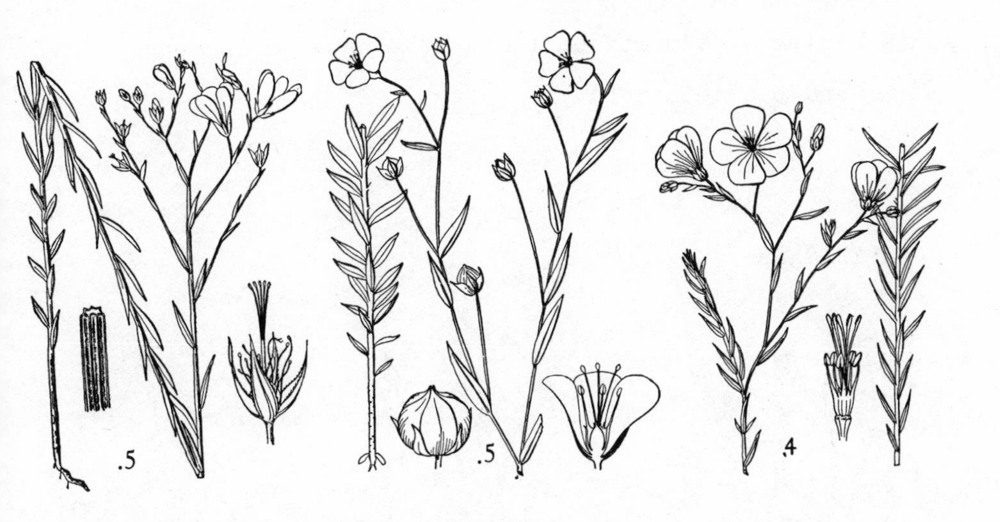

Linum sulcatum	Linum usitatissimum	Linum lewisii
-	-	-
flax	flax	flax

LINACEAE

dHypoG/(4-)5L(4-)5,10,A_(4-)5C(4-)5K(4-)5/caps.,bac.

Dicotyledonous plants with hypogynous flowers. Gynoecium

of 5 united carpels (or 4). Locules 5, or 10 (or 4). Androecium with parts united at the base, of 5 stamens (or 4). Corolla of 5 petals (or 4). Calyx of 5 sepals (or 4). Fruit a capsule, or baccate.

Linum sp.

dHypoGi̸5L10A_5(+S5)C5K5/

Androecium with parts united at the base, 5 stamens (and 5 staminodes).

OXALIDACEAE

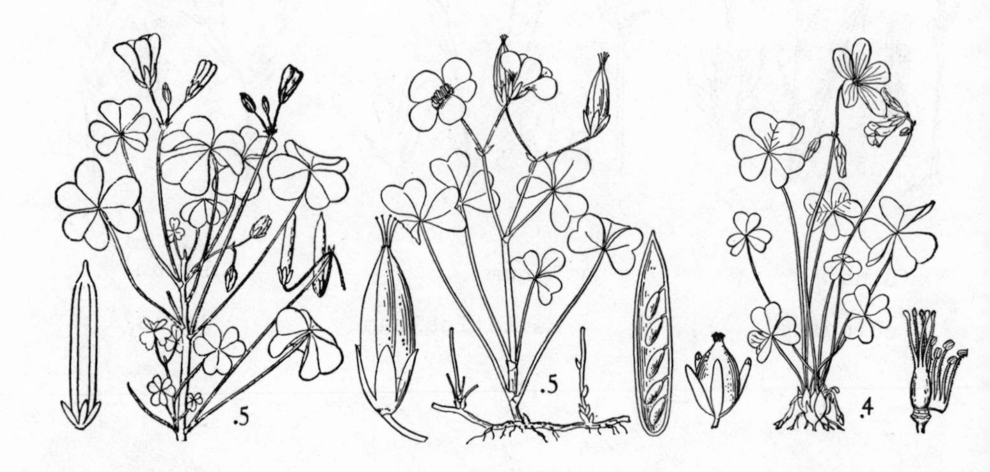

Oxalis stricta	Oxalis corniculata	Oxalis violacea
-	-	-

OXALIDACEAE

dHypoGi̸5A_10Ci,_,5K5/caps.,bac.,lvs.trifol.

Dicotyledonous plants with hypogynous flowers. Gynoecium of 5 united carpels. Androecium of 10 stamens united at the base. Corolla of 5 petals, free or united at the base. Calyx of 5 sepals. Fruit a capsule, or baccate. Leaves trifoliate.

Oxalis sp.

dHypoGi̸5A_10C_5K5/

GERANIACEAE

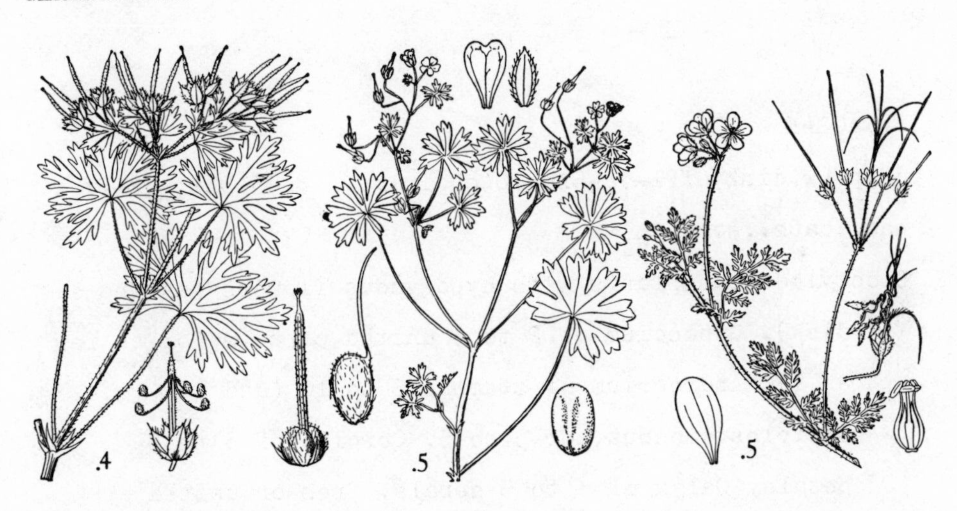

Geranium carolinianum

-

Geranium pusillum

-

Erodium cicutarium

-

GERANIACEAE

dHypoG⁄5Ai,_,10(obdipl.),5,Ca,⁄a,5K5/caps.w.beak

Dicotyledonous plants with hypogynous flowers. Gynoecium of 5 united carpels. Androecium of stamens free, or united at the base, 10 (and obdiplostemonous), or 5. Corolla of 5 petals, actinomorphic or zygomorphic. Calyx of 5 sepals. Fruit a capsule with a beaked apex.

Geranium sp.

dHypoG⁄5A10(obdipl.)S5(glan.)C5K5/

Androecium of 10 stamens (obdiplostemonous). Staminodia 5 (glandular).

RUTACEAE

dHypo(w.disk)G⁄(1-)2-5A6-10(obdipl.),3-5,C3-5Ki,_,3-5/
bac.,caps.,sam.

Dicotyledonous plants with hypogynous flowers (having a disk). Gynoecium of 2 to 5 united carpels (or as few as 1). Androecium of stamens 6 to 10 (and obdiplostemonous, or 3 to 5. Corolla of 3 to 5 petals. Calyx of 3 to 5 sepals, free or united at the base. Fruit baccate, or a capsule, or a samara.

Ruta graveolens

dHypo(w.disk)G⁄5A10(obdipl.)C5K5/

RUTACEAE

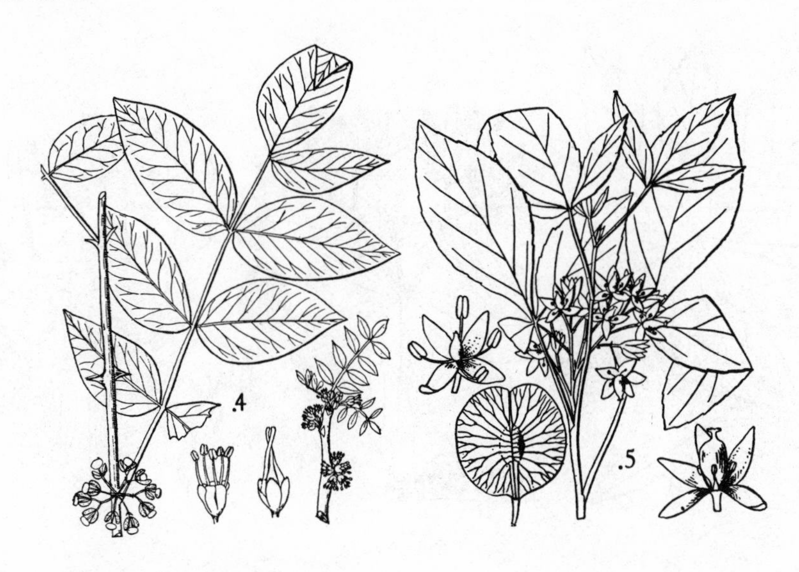

Zanthoxylum americanum	Ptelea trifoliata
-	wafer ash
prickly ash	-

EUPHORBIACEAE

dHypo(w.,not w.,disk),Psilo,G⁄(2-)3L(1-)3A1-nC4-6,0,

K,P,i,_,4-6,0/monoec.,dioec.,caps.(schizoc.)

Dicotyledonous plants with hypogynous flowers (having, or lacking, a disk), or with psilogynous flowers. Gynoecium of 3 united carpels (or as few as 2) and with 3 locules (or as few as 1). Androecium of 1

EUPHORBIACEAE

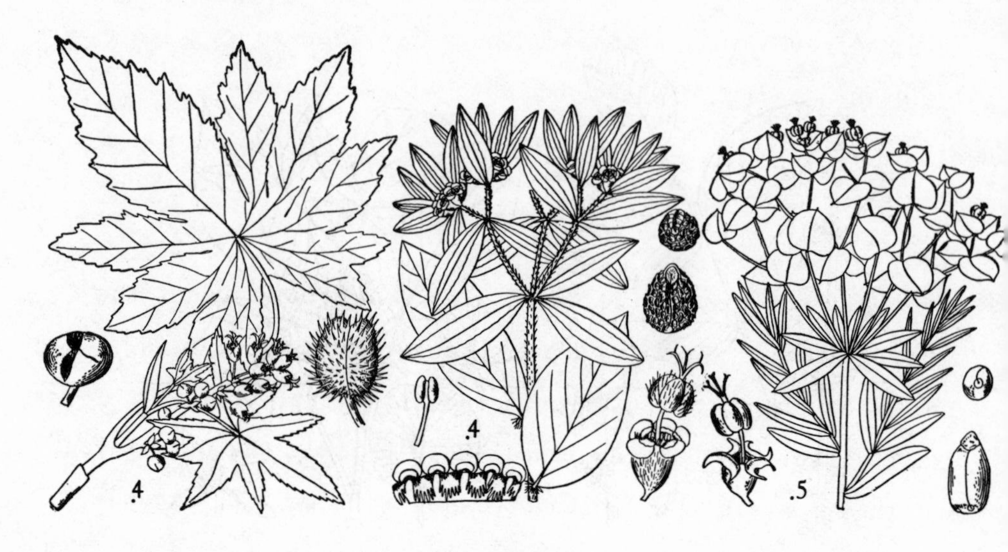

Ricinus communis — castor bean

Euphorbia marginata — snow-on-the-mountain

Euphorbia esula — leafy spurge

to many stamens. Corolla of 4 to 6 petals, or none. Calyx, or perianth, of sepals, or tepals, free or united at the base, 4 to 6, or none. Flowers unisexual and the plants monoecious, or dioecious. Fruit a capsule (schizocarp).

Euphorbia sp.

dHypo(w.gynoph.(f))Gi̸3A1/monoec.pseudanth.(cyath.)

juice milky

Dicotyledonous plants with hypogynous flowers (with gynophore (in pistillate flowers)). Flowers aggregated into pseudanthia (cyathia). Juice milky.

25 3:1

Endlicher places the Geraniales toward the end of the Polypetalidae, the third of his three dicot subclasses. The order contains the Geraniaceae, Linaceae, and Oxalidaceae. The Rutales immediately precede the Geraniales, and they contain the Juglandaceae as well as the Rutaceae. The Euphorbiales immediately precede the Rutales and of the families here considered contain only the Euphorbiaceae. Bentham and Hooker place the Geraniales toward the middle of the Polypetalidae, the first of their three dicot subclasses. Here the Geraniales are the first of an order group of plants with flowers having disks. Their order contains the Linaceae, Geraniaceae (incl. Oxalis), and Rutaceae. The Euphorbiaceae are included in the Urticales of the subclass Apetalidae. Eichler places the Geraniales toward the middle of the Agamopetalidae, the second of his two dicot subclasses. Here the Geraniales are the first of an order group of plants with basically eucyclic flowers. The order contains the Geraniaceae, Oxalidaceae, Linaceae, and Rutaceae. The Euphorbiaceae immediately follow the Geraniaceae and of the families here considered contain only the Euphorbiaceae. Engler places the Geraniales within the polypetalous portion of the Agamopetalidae, the first of his two dicot subclasses. Here the Geraniales with the Sapindales and Rhamnales form a basically eucyclic order group. The Geraniales contain the Linaceae, Oxalidaceae, Geraniaceae, Rutaceae, and Euphorbiaceae. The species of the Euphorbiaceae occurring in our area seem anomalous in the Geraniales. Bessey places the Geraniales toward the beginning of the Dicotostrobiloididae, the first of his two dicot subclasses. He diagrams as a phylogenetic sequence: Magnoliales, Geraniales. The Besseyan Geraniales contain the Geraniaceae, Oxalidaceae, Linaceae, Rutaceae, and Euphorbiaceae. Hutchinson places the Geraniales toward the end of the Herbacidae, the second of his two dicot subclasses. He diagrams as a phylogenetic sequence: Saxifragales, Geraniales. The Hutchinsonian Geraniales contain the Geraniaceae and Oxalidaceae. The Linaceae, Rutaceae, and Euphorbiaceae are, however, classified in separate orders of the subclass Lignosidae. Melchior maintains the Geraniales in approximately their Englerian position. The order includes the Oxalidaceae, Geraniaceae, Linaceae, and Euphorbiaceae. The Rutales immediately follow the Geraniales, and of the families here considered they contain only the Rutaceae. Cronquist places the Geraniales in the Rosidae, the fifth of his six dicot subclasses. He diagrams as a phylogenetic sequence: Rosales, Sapindales, Geraniales. The order Geraniales includes the Oxalidaceae and Geraniaceae. The Linales immediately follow the Geraniales and of the families here considered contain only the Linaceae. The Rutaceae are assigned to the Sapindales, which immediately precede the Geraniales. The Euphorbiales follow the Celastrales elsewhere in the subclass Rosidae, and this sequence is also presented in a phylogenetic diagram.

25 4:1

GLOSSARY FOR GERANIALES

Linales. Order in subclass Agamopetalidae.

Malpighiales. Order in division Lignosae (Hutch.)

Linales. Order in subclass Rosidae (Cronq.)

Geraniales. Order in subclass Agamopetalidae.

Gruinales. Classis in cohors Dialypetalae (Endl.)

Geraniales. Cohors in series Disciflorae (B.&H.) In subclassis Polypetalae.

Gruinales. Unterreihe in Reihe Eucyclicae (Eichl.) In Klasse Chori- und Apetalae.

Geraniales. Series in subclassis Archichlamydeae (Engl.)

Geraniales. Order in superorder Apopetalae-Polycarpellatae (Bessey). In subclass Strobiloideae.

Geraniales. Order in division Herbaceae (Hutch.)

Geraniales. Reihe in Unterklasse Archichlamydeae (Melch.)

Geraniales. Order in subclass Rosidae (Cronq.)

Rutales. Order in subclass Agamopetalidae.

Terebinthineae. Classis in cohors Dialypetalae (Endl.)

Terebinthinae. Unterreihe in Reihe Eucyclicae (Eichl.)

In Klasse Chori- und Apetalae.

Rutales. Order in division Lignosae (Hutch.)

Rutales. Reihe in Unterklasse Archichlamydeae (Melch.)

Euphorbiales. Order in subclass Agamopetalidae.

Tricoccae. Classis in cohors Dialypetalae (Endl.)

Tricoccae. Unterreihe in Reihe Tricoccae (sic) (Eichl.)

In Klasse Chori- und Apetalae.

Euphorbiales. Order in division Lignosae (Hutch.)

Euphorbiales. Order in subclass Rosidae (Cronq.)

Linaceae. Family in order Geraniales.

Lineae. Ordo in classis Gruinales (Endl.)

Lineae. Ordo in cohors Geraniales (B.&H.)

Linaceae. Familie in Unterreihe Gruinales (Eichl.)

Linaceae. Familia in series Geraniales (Engl.)

Linaceae. Family in order Geraniales (Bessey).

Linaceae. Family in order Malpighiales (Hutch.)

Linaceae. Familie in Reihe Geraniales (Melch.)

Linaceae. Family in order Linales (Cronq.)

Oxalidaceae. Family in order Geraniales.

Oxalideae. Ordo in classis Gruinales (Endl.)

Oxalideae. Familie in Unterreihe Gruinales (Eichl.)

Oxalidaceae. Familia in series Geraniales (Engl.)

Oxalidaceae. Family in order Geraniales (Bessey), (Hutch.), (Cronq.)

Oxalidaceae. Familie in Reihe Geraniales (Melch.)

Geraniaceae. Family in order Geraniales.

Geraniaceae. Ordo in classis Gruinales (Endl.)

Geraniaceae. Ordo in cohors Geraniales (B.&H.)

Geraniaceae. Familie in Unterreihe Gruinales (Eichl.)

Geraniaceae. Familia in series Geraniales (Engl.)

Geraniaceae. Family in order Geraniales (Bessey), (Hutch.), (Cronq.)

Geraniaceae. Familie in Reihe Geraniales (Melch.)

Rutaceae. Family in order Geraniales.

Rutaceae. Ordo in classis Terebinthineae (Endl.)

Rutaceae. Ordo in cohors Geraniales (B.&H.)

Rutaceae. Familie in Unterreihe Terebinthinae (Eichl.)

Rutaceae. Familia in series Geraniales (Engl.)

Rutaceae. Family in order Geraniales (Bessey).

Rutaceae. Family in order Rutales (Hutch.)

Rutaceae. Familie in Reihe Rutales (Melch.)

Rutaceae. Family in order Sapindales (Cronq.)

Euphorbiaceae. Family in order Geraniales.

Euphorbiaceae. Ordo in classis Tricoccae (Endl.)

Euphorbiaceae. Ordo in series Unisexuales (B.&H.)

Euphorbiaceae. Familie in Unterreihe Tricoccae (Eichl.)

Euphorbiaceae. Familia in series Geraniales (Engl.)

Euphorbiaceae. Family in order Geraniales (Bessey).

Euphorbiaceae. Family in order Euphorbiales (Hutch.), (Cronq.)

Euphorbiaceae. Familie in Reihe Geraniales (Melch.)

Juglandaceae. See 13 4.

Juglandeae. Ordo in classis Terebinthineae (Endl.)

Balsaminaceae. See 26 4.

Balsamineae. Ordo in classis Gruinales (Endl.)

Balsaminaceae. Familie in Unterreihe Gruinales (Eichl.)

Balsaminaceae. Family in order Geraniales (Bessey), (Hutch.), (Cronq.)

26 1:1

SAPINDALES

The Sapindales include the families Aquifoliaceae, Celastraceae, Aceraceae, Sapindaceae, and Balsaminaceae. The Aquifoliaceae have coriaceous leaves, evergreen or deciduous. They lack the disk generally characteristic in the order. The Celastraceae have a disk well developed, and the gynoecium is commonly embedded in it. The Aceraceae have gynoecia with two carpels developing into a schizocarp separating into two samaras. The Sapindaceae have a hypogynous disk. Their leaves are compound and alternate. The Balsaminaceae have strongly zygomorphic flowers with connation of anthers around the stigma.

The name Aquifoliaceae is based on the generic name Aquifolium applied to the taxon now known as Ilex aquifolium, English ivy. Aquifolium is an ancient Latin name for holly and signifies etymologically: "plant with pointed leaves." Celastrus is derived from the ancient Greek name for the taxon now known as Ilex aquifolium, but its literal signification is obscure. In any case since the Ilex taxon had already been named, Celastrus in botanical nomenclature was applied to a related one. Acer is the ancient Latin name for the maple. The ultimate signification of the name is unknown. Sapindus is a coined name based on a combination of the Latin sapo, "soap," and Indicus, (South American) "Indian," and alludes to a soap prepared from the berries of S. saponaria. Sacrificed to euphony in this formation is philology, which might suggest such a form as: Saponindicus. Balsaminaceae is based on an old generic name Balsamina once applied to the plant now known as Impatiens balsamina. Balsamina is a feminine noun derived from the ancient Latin balsaminus, "of balsam," formed on a word referring to the plant now known as Amytis opobalsamum, a rutaceous tree named in ancient Latin:

<u>balsamum</u>. From this last was derived the potent balsam or balm once used in embalming. The ultimate origin of the name is obscure, but it has Greek antecedents that are interpreted as having Semitic origin.

26 2:1

AQUIFOLIACEAE

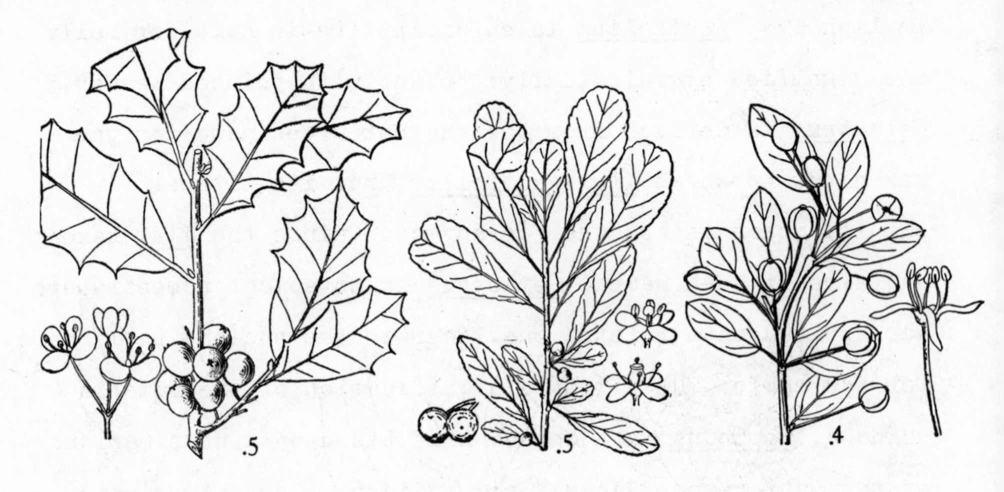

Ilex opaca	Ilex decidua	Nemopanthus mucronatus
American holly	-	-
		mountain holly

AQUIFOLIACEAE

dHypoG/4-8A4-8Ci,_,4-8K_4-6/drup.evg.

Dicotyledonous plants with hypogynous flowers. Gynoecium of 4 to 8 united carpels. Androecium of 4 to 8 stamens. Corolla of 4 to 8 petals, free or united at the base. Calyx of 4 to 6 sepals united at the base. Fruit drupaceous. Plants evergreen.

Ilex sp.

dHypoG/4A4C4K_4/

CELASTRACEAE

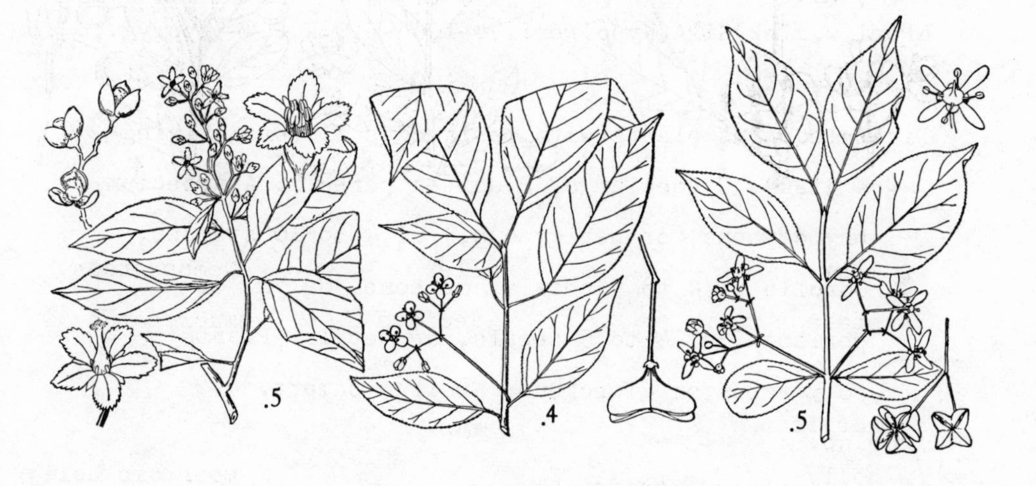

Celastrus scandens	Euonymus atropurpureus	Euonymus europaeus
-	burning bush	spindle tree
bittersweet	-	-

CELASTRACEAE

dHypo,Peri,(w.disk)G/2-5A4-5C4-5K_4-5/caps.lvs.simp.

Dicotyledonous plants with hypogynous, or perigynous, flowers (having a disk). Gynoecium of 2 to 5 united carpels. Androecium of 4 to 5 stamens. Corolla of 4 to 5 petals. Calyx of 4 to 5 sepals united at the base. Fruit a capsule. Leaves simple.

Euonymous sp.

dHypo(w.disk)G/3A5C5K_5/

ACERACEAE

dPeri(w.disk)G/2A(Hypo,Peri,)4-10C4-5,0,K,P,4-5/polyg., dioec.,sam.

Dicotyledonous plants with perigynous flowers (having a disk). Gynoecium of 2 united carpels. Androecium (hypogynous, or perigynous,) of 4 to 10 stamens. Corolla of 4 to 5 petals, or none. Calyx, or perianth, of 4 to 5 sepals, or tepals. Plants polygamous or dioecious. Fruit a samara.

Acer sp.

dPeri(w.disk)G/2A8C5K5/polyg.

ACERACEAE

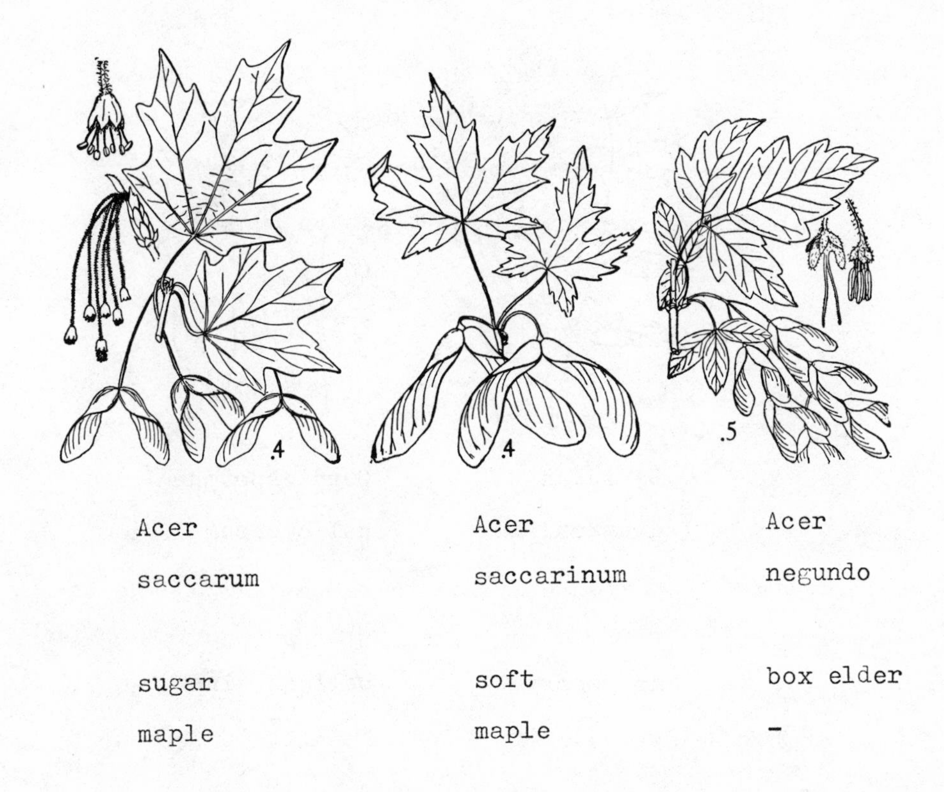

SAPINDACEAE

dHypo(w.disk extrastam.)G/3A(5-)8(-10)Cø,a,5K5/perf.,
polyg.-dioec.,caps.,drup.,bac.,lvs.comp.

Dicotyledonous plants with hypogynous flowers (having an extrastaminal disk). Gynoecium of 3 united carpels. Androecium of 8 stamens (or as few as 5, or as many as 10). Corolla zygomorphic, or actinomorphic, of 5 petals. Calyx of 5 sepals. Flowers perfect, or plants

SAPINDACEAE

Sapindus drummondii	Cardiospermum halicacabum
-	-
soapberry	balloon vine

polygamodioecious. Fruit a capsule, or a drupe, or baccate. Leaves compound.

Koelreuteria paniculata

dHypo(w.disk extrastam.)G/3A8C/4K5/polyg.-dioec.

BALSAMINACEAE

dHypoG/5A‾5C/5(1+/2+/2)K/3(-5)(lw.spur)/caps.,bac.

Dicotyledonous plants with hypogynous flowers. Gynoecium

BALSAMINACEAE

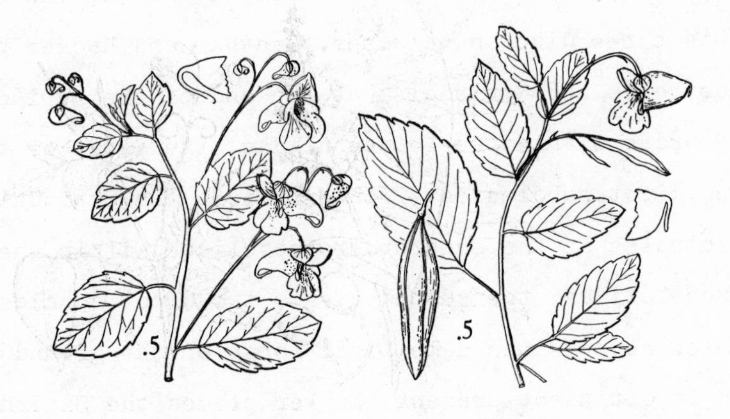

Impatiens biflora	Impatiens pallida
-	-
touch-me-not	touch-me-not

of 5 united carpels. Androecium of 5 stamens united by the anthers. Corolla zygomorphic of 5 petals (a single petal and two sets of 2 united petals. Calyx zygomorphic of 3 sepals (or as many as 5) (1 with a spur). Fruit a capsule, or baccate.

Impatiens sp.

dHypoG1̸5A¯5Ca̸(1+1̸2+1̸2)Ka̸3(lw.spur)/

Endlicher places the Sapindales and Celastrales in an unnamed order group within the Polypetalidae, the third of his three dicot subclasses. Bentham and Hooker place the Celastrales and Sapindales at the end of their Disciflorae superorder within the Polypetalidae, the first of their three dicot subclasses. Eichler places the Sapindales and Celastrales in the superorder Eucyclicae within the Agamopetalidae, the second of his two dicot subclasses. Eichler rejects the name Disciflorae on the ground that a disk is often not present. Engler places the Sapindales in a eucyclic order group within the Agamopetalidae, the first of his two dicot subclasses. He alone includes the Celastraceae within the Sapindales. Bessey places the Celastrales and Sapindales in the superorder Apopetalae within the Dicotocotyloididae, the second of his two dicot subclasses. He diagrams as a phylogenetic sequence: Magnoliales, Rosales, Celastrales, Sapindales. Hutchinson places the Celastrales remote from the Sapindales, although both orders stand within the Lignosidae, the first of his two dicot subclasses. He diagrams the Celastrales as derived from the Magnoliales, and he diagrams the Sapindales as derived indirectly from the Rutales and ultimately from the Magnoliales. Melchior retains the Sapindales in their Englerian position, but he splits off the Celastraceae and allies and raises them to ordinal rank. Cronquist places the Sapindales in a position separated from the Celastrales, although both orders stand within the Rosidae, the fifth of his six dicot subclasses. He diagrams as a phylogenetic sequence: Rosales, Sapindales. He also diagrams another line from the Rosales, which branches to give rise to both the Celastrales and the Rhamnales.

Thus the Sapindales and Celastrales may be combined into a large Sapindales order (Engler), or they may be kept distinct. When distinct, the orders may be adjacent or separated. When adjacent, the Sapindales may be in first position (Eichler, Melchior) or in second position (Bentham and Hooker, Bessey). When separated, the orders may be relatively close (Endlicher, Cronquist), or they may be remote from each other (Hutchinson).

The families of Endlicher's Sapindales have the sequence: Aceraceae, Sapindaceae. Melchior concurs in the sequence. Eichler, however, adopts the sequence: Sapindaceae, Aceraceae. Bessey, Hutchinson, and Cronquist concur in this sequence. Engler follows Endlicher with respect to the sequence of these families, but he also includes in the order the Aquifoliaceae and Celastraceae. The latter family is more usually assigned to the Celastrales, as has been noted. He and Melchior include here the Balsaminaceae, a family more usually assigned to the Geraniales. Bessey alone introduces apetalous amentiferous families into the Sapindales. His sequence runs: Sapindaceae, Aceraceae, Juglandaceae, Betulaceae, Fagaceae. Cronquist alone recognizes the

Rutales as members of the order. His sequence runs: Sapindaceae, Aceraceae, Rutaceae. Bentham and Hooker do not recognize the Aceraceae but include Acer within the Sapindaceae.

The families of Endlicher's Celastrales have the sequence: Celastraceae, Aquifoliaceae, Rhamnaceae. Eichler extends this to Celastraceae, Aquifoliaceae, Rhamnaceae, Vitaceae. Bessey adjusts this to Rhamnaceae, Vitaceae, Celastraceae, Aquifoliaceae. Following these families he also includes the Elaeagnaceae and Santalaceae. Hutchinson, Melchior, and Cronquist exclude the Rhamnaceae and Vitaceae from the Celastrales. Hutchinson and Melchior have the sequence Aquifoliaceae, Celastraceae. Cronquist has the sequence: Celastraceae, Aquifoliaceae. Bentham and Hooker have the sequence: Celastraceae, Rhamnaceae, Vitaceae. These authors alone place the Aquifoliaceae in an order separate from that to which they assign the Celastraceae.

There is then a consensus that the Sapindaceae and Aceraceae are related but sufficiently distinct to constitute separate families. There is also a tendency to place the Sapindaceae ahead of the Aceraceae. The order Celastrales is usually regarded as related to the Sapindales, but modern practice excludes the Rhamnaceae and Vitaceae from the order. There is consensus that the Celastraceae and Aquifoliaceae belong in the order, but the consensus does not extend to their sequence within the order.

GLOSSARY FOR SAPINDALES

Aquifoliales. Order in subclass Agamopetalidae.

Olacales. Cohors in series Disciflorae (B.&H.)

In subclassis Polypetalae.

Celastrales. Order in subclass Agamopetalidae.

Frangulaceae. Classis in cohors Dialypetalae (Endl.)

Celastrales. Cohors in series Disciflorae (B.&H.)

In subclassis Polypetalae.

Frangulinae. Unterreihe in Reihe Eucyclicae (Eichl.)

In Klasse Chori- und Apetalae.

Celastrales. Order in superorder Apopetalae (Bessey)

In subclass Cotyloideae.

Celastrales. Order in division Lignosae (Hutch.)

Celastrales. Reihe in Unterklasse Archichlamydeae (Melch.)

Celastrales. Order in subclass Rosidae (Cronq.)

Sapindales. Order in subclass Agamopetalidae.

Acera. Classis in cohors Dialypetalae (Endl.)

Sapindales. Cohors in series Disciflorae (B.&H.)

In subclass Polypetalae.

Aesculinae. Unterreihe in Reihe Eucyclicae (Eichl.)

In Klasse Chori- und Apetalae.

Sapindales. Series in subclassis Archichlamydeae (Engl.)

Sapindales. Order in superorder Apopetalae (Bessey).

In subclass Cotyloideae.

Sapindales. Order in division Lignosae (Hutch.)

Sapindales. Reihe in Unterklasse Archichlamydeae (Melch.)

Sapindales. Order in subclass Rosidae (Cronq.)

Aquifoliaceae. Family in order Sapindales.

Ilicineae. Ordo in classis Frangulaceae (Endl.)

Ilicineae. Ordo in cohors Olacales (B.&H.)

Aquifoliaceae. Familie in Unterreihe Frangulinae (Eichl.)

Aquifoliaceae. Familia in series Sapindales (Engl.)

Aquifoliaceae. Family in order Celastrales (Bessey), (Hutch.), (Cronq.)

Aquifoliaceae. Familie in Reihe Celastrales (Melch.)

Celastraceae. Family in order Sapindales.

Celastrineae. Ordo in classis Frangulaceae (Endl.)

Celastrineae. Ordo in cohors Celastrales (B.&H.)

Celastraceae. Familie in Unterreihe Frangulinae (Eichl.)

Celastraceae. Familia in series Sapindales (Engl.)

Celastraceae. Family in order Celastrales (Bessey), (Hutch.), (Cronq.)

Celastraceae. Familie in Reihe Celastrales (Melch.)

Aceraceae. Family in order Sapindales.

Acerineae. Ordo in classis Acera (Endl.)

Acerineae. Familie in Unterreihe Aesculinae (Eichl.)

Aceraceae. Familia in series Sapindales (Engl.)

Aceraceae. Family in order Sapindales (Bessey), (Hutch.), (Cronq.)

Aceraceae. Familie in Reihe Sapindales (Melch.)

Sapindaceae. Family in order Sapindales.

Sapindaceae. Ordo in classis Acera (Endl.)

Sapindaceae. Ordo in cohors Sapindales (B.&H.)

Sapindaceae. Familie in Unterreihe Aesculinae (Eichl.)

Sapindaceae. Familia in series Sapindales (Engl.)

Sapindaceae. Family in order Sapindales (Bessey), (Hutch.), (Cronq.)

Sapindaceae. Familie in Reihe Sapindales (Melch.)

Balsaminaceae. Family in order Sapindales.

Balsamineae. Ordo in classis Gruinales (Endl.)

Balsaminaceae. Familie in Unterreihe Gruinales (Eichl.)

Balsaminaceae. Familia in series Sapindales (Engl.)

Balsaminaceae. Family in order Geraniales (Bessey), (Hutch.), (Cronq.)

Balsaminaceae. Familie in Reihe Sapindales (Melch.)

Myricaceae. See 11 4.

Myricaceae. Family in order Sapindales (Bessey).

Juglandaceae. See 13 4.

Juglandaceae. Family in order Sapindales (Bessey).

Betulaceae. See 14 4.

Betulaceae. Family in order Sapindales (Bessey).

Fagaceae. See 14 4.

Fagaceae. Family in order Sapindales (Bessey).

Santalaceae. See 16 4.

Santalaceae. Family in order Celastrales (Bessey).

Rutaceae. See 25 4.

Rutaceae. Family in order Sapindales (Cronq.)

Rhamnaceae. See 27 4.

Rhamneae. Ordo in classis Frangulaceae (Endl.)

Rhamneae. Ordo in cohors Celastrales (B.&H.)

Rhamnaceae. Familie in Unterreihe Frangulinae (Eichl.)

Rhamnaceae. Family in order Celastrales (Bessey).

Vitaceae. See 27 4.

Ampelideae. Ordo in cohors Celastrales (B.&H.)

Ampelideae. Familie in Unterreihe Frangulinae (Eichl.)

Vitaceae. Family in order Celastrales (Bessey).

Elaeagnaceae. See 31 4.

Elaeagnaceae. Family in order Celastrales (Bessey).

27 1:1

RHAMNALES

The Rhamnales include the families Rhamnaceae and Vitaceae. The position of the stamens opposite the petals, both groups on a hypanthium, is a striking feature of _Rhamnus_. The Vitaceae have an intrastaminal disk.

The name _Rhamnus_ is derived through Latin from an ancient Greek word applied to various thorny shrubs. _Vitis_ is an ancient Latin name signifying "vine."

RHAMNACEAE

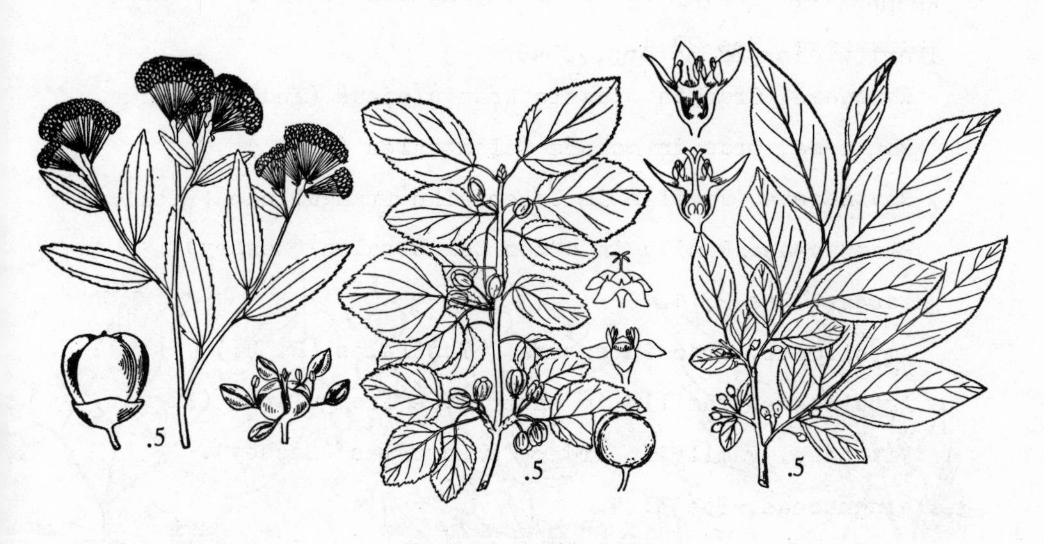

Ceanothus	Rhamnus	Rhamnus
ovatus	cathartica	lanceolata
-	-	-
	buckthorn	buckthorn

RHAMNACEAE

dPeri,Epi,(w.disk)G/2-4A4-5(opp.)C4-5K4-5/perf.,

polyg.-dioec.,drup.,caps.

Dicotyledonous plants with perigynous, or epigynous, flowers (having a disk). Gynoecium of 2 to 4 united carpels. Androecium of 4 to 5 stamens (opposite the petals). Corolla of 4 to 5 petals. Calyx of 4 to 5 sepals. Flowers perfect, or plants polygamodioecious.

Fruit drupaceous, or a capsule.

Rhamnus sp.

dPeri(w.disk)G/3A5(opp.)C5K5/

VITACEAE

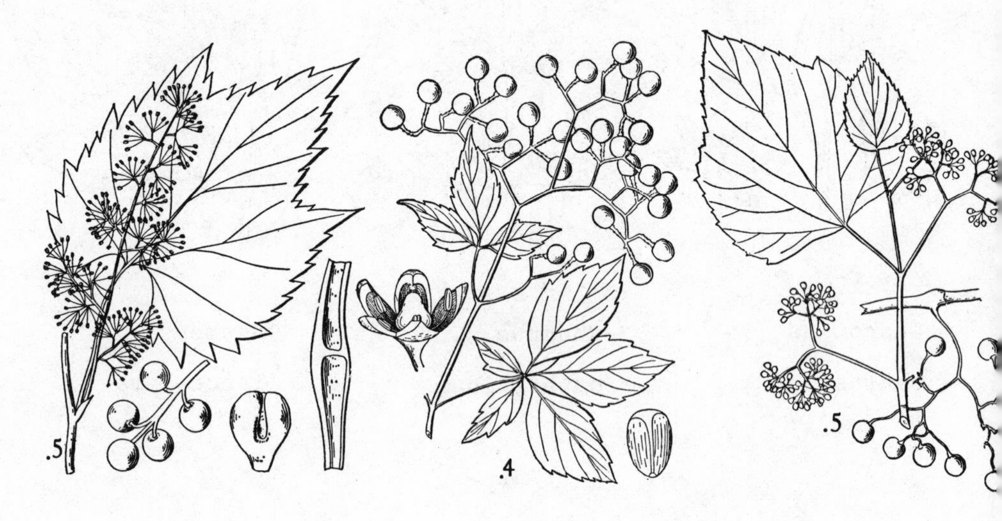

Vitis vulpina	Parthenocissus quinquefolia	Ampelopsis cordata
wild grape	- woodbine	-

VITACEAE

dHypo,Peri,(w.disk)G/2A4-5(opp.)C,P,i,¯,4-5(valv.)K(absent),(entire),/4-5/perf.,unisex.,ber.

RHAMNALES 27 2:3

Dicotyledonous plants with hypogynous, or perigynous, flowers (having a disk). Gynoecium of 2 united carpels. Androecium of 4 to 5 stamens (opposite the petals, or tepals). Corolla, or perianth, of 4 to 5 petals, or tepals, (valvate) free, or united at the apex. Calyx absent, or entire, or of 4 to 5 united sepals. Flowers perfect, or unisexual. Fruit a berry.

Vitis sp.

dPeri(w.disk)G/2A5(opp.)P¯5(valv.)/

27 3:1

Endlicher, Bentham and Hooker, Eichler, and Bessey do not recognize the Rhamnales. They place the Rhamnaceae in the Celastrales. For their treatment of the Celastrales see section 26. Endlicher includes the Rhamnaceae in a eucyclic order group: Sapindales, Celastrales, Euphorbiales, Rutales, Geraniales. Bentham and Hooker include the Rhamnaceae in their superorder Disciflorae: Geraniales, Aquifoliales, Celastrales, Sapindales. Eichler includes the Rhamnaceae in his superorder Eucyclicae: Geraniales, Rutales, Sapindales, Celastrales. Engler places the Rhamnales after the Sapindales in the Agamopetalidae, the first of his two dicot subclasses. He includes the Rhamnales in the basically eucyclic order group: Geraniales, Sapindales, Rhamnales. Bessey includes the Rhamnaceae in his dicotocotyloididous Apopetalae superorder wherein the Celastrales, Sapindales, and Umbellales form an order group isolated from the Geraniales. Hutchinson places the Rhamnales in the Lignosidae, the first of his two dicot subclasses. He diagrams as a phylogenetic sequence: Magnoliales, Rhamnales. Melchior maintains the Rhamnales in their Englerian position. He revises the basically eucyclic Englerian order group to run: Geraniales, Rutales, Sapindales, Celastrales, Rhamnales. Cronquist places the Rhamnales in the Rosidae, the fifth of his six dicot subclasses. He diagrams a phylogenetic line from the Rosales which divides to give rise to both the Celastrales and the Rhamnales.

Most systems associate the Rhamnaceae and Vitaceae in the same order, but Endlicher alone, while placing the Rhamnaceae in the Celastrales, assigns the Vitaceae to the Umbellales. Engler, Melchior, and Cronquist recognize the Rhamnales as containing the Rhamnaceae and Vitaceae and in this sequence. Hutchinson also recognizes this order and sequence, but he additionally includes in first position the Elaeagnaceae.

There is a modern consensus that the order Rhamnales should be recognized and associated with the Celastrales. The Rhamnales are usually regarded as consisting in our area of only two families (Hutchinson dissenting), and these families have the sequence: Rhamnaceae, Vitaceae.

GLOSSARY FOR RHAMNALES

Rhamnales. Order in subclass Agamopetalidae.

Rhamnales. Series in subclassis Archichlamydeae (Engl.)

Rhamnales. Order in division Lignosae (Hutch.)

Rhamnales. Reihe in Unterklasse Archichlamydeae (Melch.)

Rhamnales. Order in subclass Rosidae (Cronq.)

Rhamnaceae. Family in order Rhamnales.

Rhamneae. Ordo in classis Frangulaceae (Endl.)

Rhamneae. Ordo in cohors Celastrales (B.&H.)

Rhamnaceae. Familie in Unterreihe Frangulinae (Eichl.)

Rhamnaceae. Familia in series Rhamnales (Engl.)

Rhamnaceae. Family in order Celastrales (Bessey).

Rhamnaceae. Family in order Rhamnales (Hutch.), (Cronq.)

Rhamnaceae. Familie in Reihe Rhamnales (Melch.)

Vitaceae. Family in order Rhamnales.

Ampelideae. Ordo in classis Discanthae (Endl.)

Ampelideae. Ordo in cohors Celastrales (B.&H.)

Ampelideae. Familie in Unterreihe Frangulinae (Eichl.)

Vitaceae. Familia in series Rhamnales (Engl.)

Vitaceae. Family in order Celastrales (Bessey).

Vitaceae. Family in order Rhamnales (Hutch.), (Cronq.)

Vitaceae. Familie in Reihe Rhamnales (Melch.)

Elaeagnaceae. See 31 4.

Elaeagnaceae. Family in order Rhamnales (Hutch.)

MALVALES

The Malvales are an order of dicotyledons with a gynoecium of two or more carpels at least partially united into a compound ovary with axile placentation. The androecium has numerous stamens, commonly with united filaments. The corolla consists of five distinct petals. The calyx consists of five sepals characteristically valvate in the bud.

The Marvales include the Tiliaceae and Malvaceae. The Tiliaceae have stamens with two-celled anthers. The inflorescence arises upon a leaf-like bract which persists to provide the nut fruits a samara-like dissemination. The Malvaceae have monadelphous stamens with one-celled anthers.

Tilia is the ancient Latin name for the linden tree. The English name *linden* has an old Germanic root *lind* designating the tree and the adjective suffix -*en*. The philological relationship of *lind* to *linden* is the same as that of *wood* to *wooden*. The name *basswood* alludes to the bast fiber derived from the linden. *Malva* is the ancient Latin name for mallow.

TILIACEAE

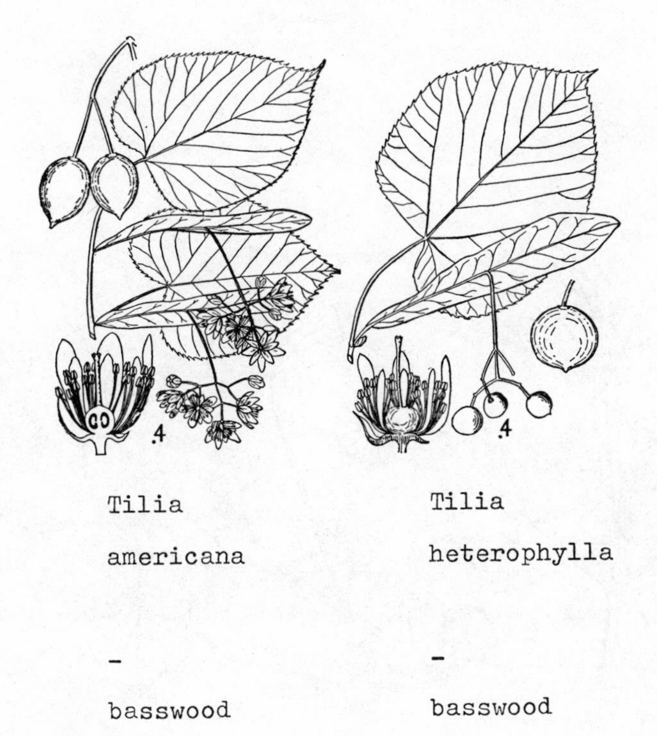

Tilia americana

-

basswood

Tilia heterophylla

-

basswood

TILIACEAE

dHypoG/5An(pentadelph.)S5,0,C5K5(valv.)/caps.,bac., infl.on lf.-like bract

Dicotyledonous plants with hypogynous flowers. Gynoecium of 5 united carpels. Androecium of numerous stamens (pentadelphous). Staminodia 5, or none. Corolla of 5 petals. Calyx of 5 sepals (valvate). Fruit a capsule, or baccate. Inflorescence on a leaf-like bract.

Tilia americana

dHypoG/5An(pentadelph.)S5C5K5(valv.)/

MALVACEAE

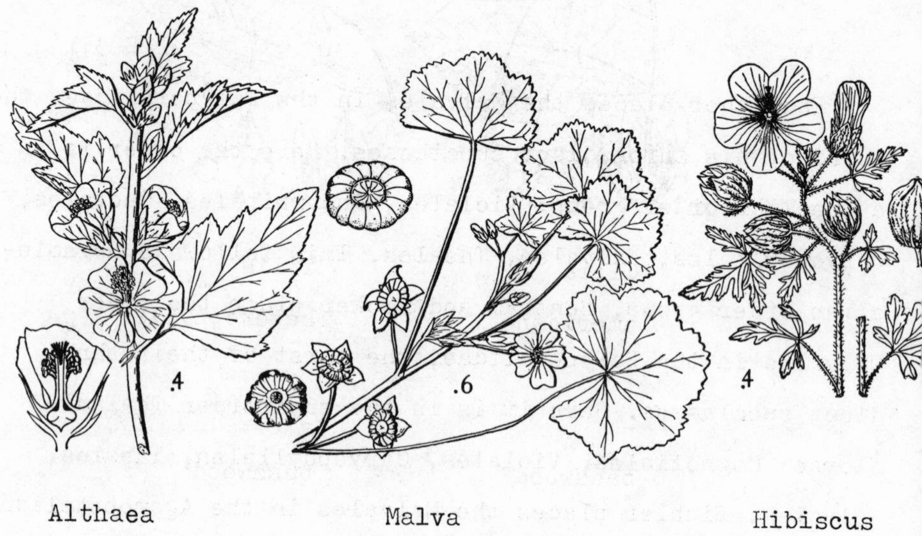

Althaea officinalis	Malva rotundifolia	Hibiscus trionum
marsh mallow	-	-
-	mallow	

MALVACEAE

dHypoG$^{\pm}$/5-nA/n(epipet.)C5K/5(valv.)/schizoc.,caps.

Dicotyledonous plants with hypogynous flowers. Gynoecium

of 5 to many more or less united carpels. Androecium of numerous united stamens (epipetalous). Corolla of 5 petals. Calyx of 5 united sepals (valvate). Fruit a schizocarp, or capsule.

Malva sp.
dHypoG/10A/n(epipet.)C5K/5(valv.)/

28 3:1

Endlicher places the Malvales in the Polypetalidae, the third of his three dicot subclasses. Here the order is in a thealian order group: Violales, Cucurbitales, Cactales, Caryophyllales, Malvales, Theales. This follows a magnolealian order group. Bentham and Hooker place the order Malvales in the Polypetalidae, the first of their three dicot subclasses. Here it is in the superorder Thalamiflorae: Magnoliales, Violales, Caryophyllales, Theales, Malvales. Eichler places the Malvales in the Agamopetalidae, the second of his two dicot subclasses. Here it is in the superorder Aphanocyclicae: Magnoliales, Papaverales, Theales, Malvales. Eichler considers that this group corresponds to that portion of the Thalamiflorae which has numerous stamens. This consideration eliminates the Caryophyllales. Engler places the Malvales in the Agamopetalidae, the first of his two dicot subclasses. He lets the Malvales fall in the order group: Malvales, Theales. In later versions of his system, however, these orders fall in separate order groups. Bessey places the Malvales in the Dicotostrobiloididae, the first of his two dicot subclasses. Here they are in the superorder Apopetalae-Polycarpellatae. Bessey diagrams as a phylogenetic sequence: Magnoliales, Malvales. Hutchinson places the Malvales in the Lignosidae, the first of his two dicot subclasses. He diagrams as a phylogenetic sequence: Tiliales, Malvales. Melchior maintains the Malvales in approximately their Englerian position. He places them in a violalian order group: Malvales, Elaeagnales, Violales, Cucurbitales. Cronquist places the Malvales in the Theidae, the fourth of his six dicot subclasses. He diagrams as a phylogenetic sequence: Theales, Malvales.

In all the systems except that of Hutchinson the Malvales include the Tiliaceae as well as the Malvaceae. Hutchinson, however, splits off the Tiliaceae and raises them to ordinal rank. Bessey alone includes the families Ulmaceae, Moraceae, and Urticaceae within the Malvales.

Within this order, Endlicher recognizes the sequence: Malvaceae, Tiliaceae. Only Bessey concurs on this point, and his position is complicated by his inclusion of the urticalian families within the order. Bentham and Hooker accept the opposite sequence: Tiliaceae, Malvaceae. Eichler, Engler, Melchior, and Cronquist concur in this; and Hutchinson places his Tiliales before his Malvales.

There appears consensus that the Malvales include the

Tiliaceae and Malvaceae (Hutchinson dissenting), and that among the families here considered the order consists of only these two families (Bessey dissenting). There is also consensus that the families have the sequence: Tiliaceae, Malvaceae.

GLOSSARY FOR MALVALES

Tiliales. Order in subclass Agamopetalidae.

  Tiliales. Order in division Lignosae (Hutch.)

Malvales. Order in subclass Agamopetalidae.

  Columniferae. Classis in cohors Dialypetalae (Endl.)

  Malvales. Cohors in series Thalamiflorae (B.&H.)

    In subclassis Polypetalae.

  Columniferae. Unterreihe in Reihe Aphanocyclicae (Eichl.)

    In Klasse Chori- und Apetalae.

  Malvales. Series in subclassis Archichlamydeae (Engl.)

  Malvales. Order in superorder Apopetalae-Polycarpellatae

    (Bessey). In subclass Strobiloideae.

  Malvales. Order in division Lignosae (Hutch.)

  Malvales. Reihe in Unterklasse Archichlamydeae (Melch.)

  Malvales. Order in subclass Dilleniidae (Cronq.)

Tiliaceae. Family in order Malvales.

Tiliaceae. Ordo in classis Columniferae (Endl.)

Tiliaceae. Ordo in cohors Malvales (B.&H.)

Tiliaceae. Familie in Unterreihe Columniferae (Eichl.)

Tiliaceae. Familia in series Malvales (Engl.)

Tiliaceae. Family in order Malvales (Bessey), (Cronq.)

Tiliaceae. Family in order Tiliales (Hutch.)

Tiliaceae. Familie in Reihe Malvales (Melch.)

Malvaceae. Family in order Malvales.

Malvaceae. Ordo in classis Columniferae (Endl.)

Malvaceae. Ordo in cohors Malvales (B.&H.)

Malvaceae. Familie in Unterreihe Columniferae (Eichl.)

Malvaceae. Familia in series Malvales (Engl.)

Malvaceae. Family in order Malvales (Bessey), (Hutch.), (Cronq.)

Malvaceae. Familie in Reihe Malvales (Melch.)

Ulmaceae. See 15 4.

Ulmaceae. Family in order Malvales (Bessey).

Moraceae. See 15 4.

Moraceae. Family in order Malvales (Bessey).

Urticaceae. See 15 4.

Urticaceae. Family in order Malvales (Bessey).

29 1:1

THEALES

The Theales include the families Theaceae and Violaceae. The Theaceae have actinomorphic flowers with numerous stamens. The locules of the gynoecium are isomeric with the united carpels. The Violaceae have zygomorphic, or actinomorphic flowers with five stamens. Their tricarpellary gynoecium is unilocular.

Thea is an obsolete synonym of *Camellia*. *Stewartia* is the only genus of the Theaceae represented in our area. *Viola* is the ancient Latin name for plants of the genus. The Latin word contains *-la* as diminutive suffix, but in French derivatives this suffix lost its diminutive force; so a second diminutive suffix, *-et*, was added, and both suffixes appear in the English name *violet*.

29 2:1

THEACEAE

dHypoG/3-5A±_nC5K5/caps.

Dicotyledonous plants with hypogynous flowers. Gynoecium of 3 to 5 united carpels. Androecium of many stamens more or less united at the base. Corolla of 5 petals. Calyx of 5 sepals. Fruit a capsule.

Stewartia sp.

dHypoG/5A_nC5K5/

THEACEAE

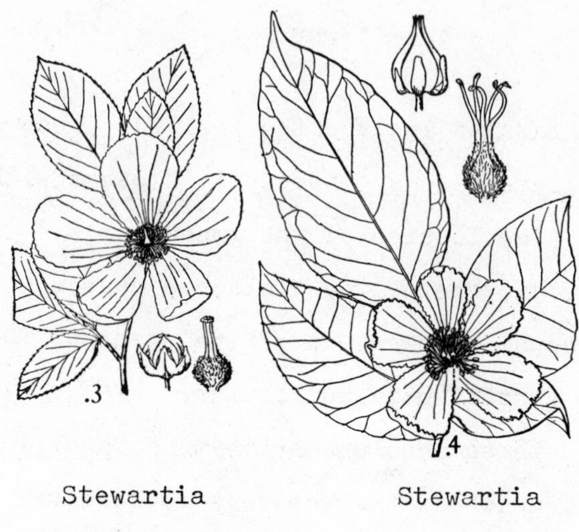

Stewartia malacodendron -

Stewartia ovata -

VIOLACEAE

dPeriG/3L1A±/5C/5(4+1 oft.w.spur)K5/caps.,ber.

Dicotyledonous plants with perigynous flowers. Gynoecium of 3 united carpels and a single locule. Androecium of 5 more or less united stamens. Corolla zygomorphic of 5 petals (in two groups, one of 4 petals and the other of a single petal often having a spur). Calyx of 5 sepals. Fruit a capsule, or a berry.

Viola sp.

VIOLACEAE

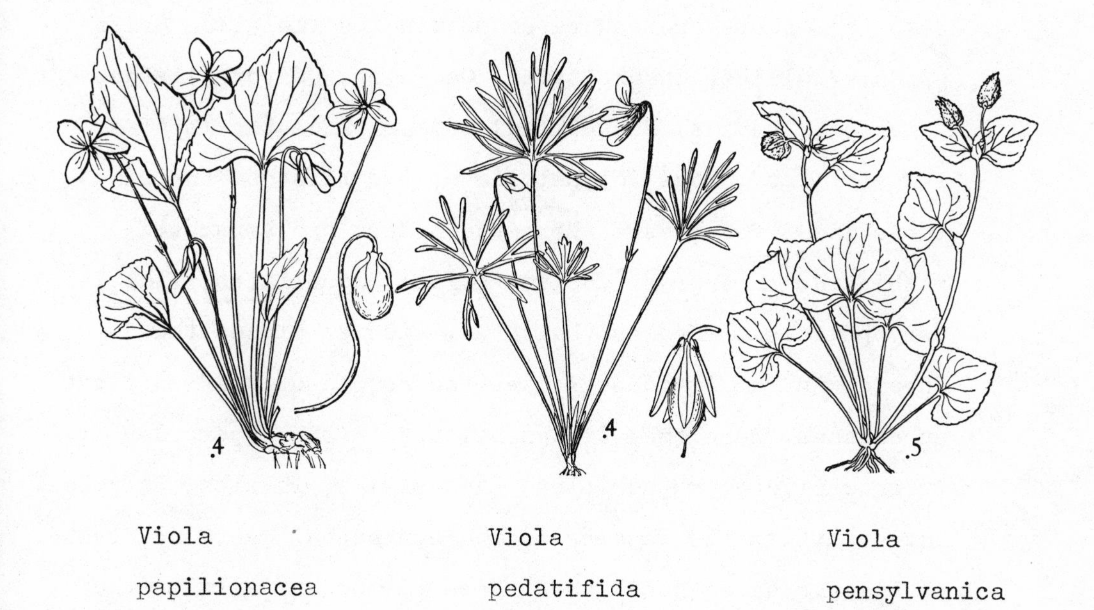

Viola papilionacea	Viola pedatifida	Viola pensylvanica
common blue violet	prairie violet	smooth yellow violet

dPeriGi̸3L1Ai̸5(2w.append.)Co̸5(4+1w.spur)K5/
Androecium of 5 united stamens (2 with appendages).

29 3:1

Endlicher, Bentham and Hooker, Hutchinson, Melchior, and Cronquist recognize both the Theales and Violales. Eichler, Engler, and Bessey include the Violaceae within the Theales.

Endlicher places the Theales and Violales in the Polypetalidae, the third of his three dicot subclasses. Here the orders are separated within the violalian order group: Violales, Cucurbitales, Cactales, Caryophyllales, Malvales, Theales. Bentham and Hooker place the Theales and Violales in the Polypetalidae, the first of their three dicot subclasses. Here the orders are separated within the superorder Thalamiflorae: Magnoliales, Violales, Caryophylales, Theales, Malvales. Eichler places the Theales in the Agamopetalidae, the second of his two dicot subclasses. Here they are associated in the superorder Aphanocyclicae: Magnoliales, Papaverales, Theales, Malvales. Engler places the Theales in the Agamopetalidae, the first of his two dicot subclasses. Here the order falls in the order group: Malvales, Theales. In later versions of his system, however, these orders fall in separate order groups. Bessey places the Theales in the Dicotostrobiloididae, the first of his two dicot subclasses. Here it falls in the superorder Apopetalae-Polycarpellatae. He diagrams as a phylogenetic sequence: Magnoliales, Theales. Hutchinson places the Theales and Violales in the Lignosidae, the first of his two dicot subclasses. Here the orders are remote from one another. Melchior places the Violales in the Agamopetalidae, the first of his two dicot subclasses. Here the Theales are associated closely with the Magnoliales. The Violales, on the other hand, retain the position of the Englerian Theales. Cronquist places the Theales and Violales in the Theidae, the fourth of his six dicot subclasses. He diagrams as a phylogenetic sequence: Theales, Violales.

Among those systems associating the Theaceae and Violaceae within a single order that of Eichler places the Violaceae far ahead of the Theaceae. The other such systems, those of Engler and Bessey, agree in placing the Theaceae ahead of the Violaceae. Endlicher includes the Droseraceae in his Violales. Bentham and Hooker include the Sarraceniaceae in that order together with the papaveralian families. Eichler includes both the Sarraceniaceae and Droseraceae in his Theales. Cronquist alone includes the Cucurbitaceae in his Violales. In these systems parietal placentation is regarded as betokening affinity among papaveralian, sarracenialian, cucurbitalian, and violalian or thealian taxa.

29 4:1

GLOSSARY FOR THEALES

Theales. Order in subclass Agamopetalidae.

Guttiferae. Classis in cohors Dialypetalae (Endl.)

Guttiferales. Cohors in series Thalamiflorae (B.&H.)
In subclassis Polypetalae.
Cistiflorae. Unterreihe in Reihe Aphanocyclicae (Eichl.)
In Klasse Chori- und Apetalae.
Parietales. Series in subclassis Archichlamydeae (Engl.)
Guttiferales. Order in superorder Apopetalae-Polycarpellatae (Bessey). In subclass Strobiloideae.
Theales. Order in division Lignosae (Hutch.)
Guttiferales. Reihe in Unterklasse Archichlamydeae (Melch.)
Theales. Order in subclass Dilleniidae (Cronq.)

Violales. Order in subclass Agamopetalidae.
Parietales. Classis in cohors Dialypetalae (Endl.)
Parietales. Cohors in series Thalamiflorae (B.&H.)
In subclassis Polypetalae.
Violales. Order in division Lignosae (Hutch.)
Violales. Reihe in Unterklasse Archichlamydeae (Melch.)
Violales. Order in subclass Dilleniidae (Cronq.)

Theaceae. Family in order Theales.
Ternstroemiaceae. Ordo in classis Guttiferae (Endl.)
Guttiferae. Ordo in cohors Guttiferales (B.&H.)
Ternstroemiaceae. Familie in Unterreihe Cistiflorae (Eichl.)
Theaceae. Familia in series Theales (Engl.)
Theaceae. Family in order Guttiferales (Bessey).
Theaceae. Family in order Theales (Hutch.), (Cronq.)
Theaceae. Familie in Reihe Guttiferales (Melch.)

Violaceae. Family in order Theales.

Violarieae. Ordo in classis Parietales (Endl.)

Violarieae. Ordo in cohors Parietales (B.&H.)

Violaceae. Familie in Unterreihe Cistiflorae (Eichl.)

Violaceae. Familia in series Parietales (Engl.)

Violaceae. Family in order Guttiferales (Bessey).

Violaceae. Family in order Violales (Hutch.), (Cronq.)

Violaceae. Familie in Reihe Violales (Melch.)

Papaveraceae. See 21 4.

Papaveraceae. Ordo in cohors Parietales (B.&H.)

Capparaceae. See 21 4.

Capparideae. Ordo in cohors Parietales (B.&H.)

Cruciferae. See 21 4.

Cruciferae. Ordo in cohors Parietales (B.&H.)

Resedaceae. See 21 4.

Resedaceae. Ordo in cohors Parietales (B.&H.)

Sarraceniaceae. See 22 4.

Sarraceniaceae. Ordo in cohors Parietales (B.&H.)

Sarraceniaceae. Familie in Unterreihe Cistiflorae (Eichl.)

Droseraceae. See 22 4.

Droseraceae. Ordo in classis Parietales (Endl.)

Droseraceae. Familie in Unterreihe Cistiflorae (Eichl.)

Cucurbitaceae. See 43 4.

Cucurbitaceae. Family in order Violales (Cronq.)

CACTALES

The Cactales contain only the family Cactaceae. The Cactaceae flowers are mostly epigynous. The ovary has three to many carpels, but is unilocular. Stamens, petals, and sepals are numerous. The vegetative body is succulent, xerophytic, and armored with spines.

Cactus is obsolete as a generic name, but it is the basis for the English name cactus applied to all plants of the family. The ancient Latin and Greek words from which the botanical name is derived applied to a thistle. The Cactaceae are plants of American origin and were unknown in ancient Greece and Rome.

Eichler omitted the Cactales from consideration in his Blüthendiagramme. He later treated the group, and his later views are here adapted to conform them to the system of his Blüthendiagramme.

30 2:1

CACTACEAE

dEpiGi̸3-nL1AnCnKn/ber.succ.(w.spines)

Dicotyledonous plants with epigynous flowers. Gynoecium of 3 to many united carpels and a single locule. Androecium of numerous stamens. Corolla of numerous

CACTACEAE

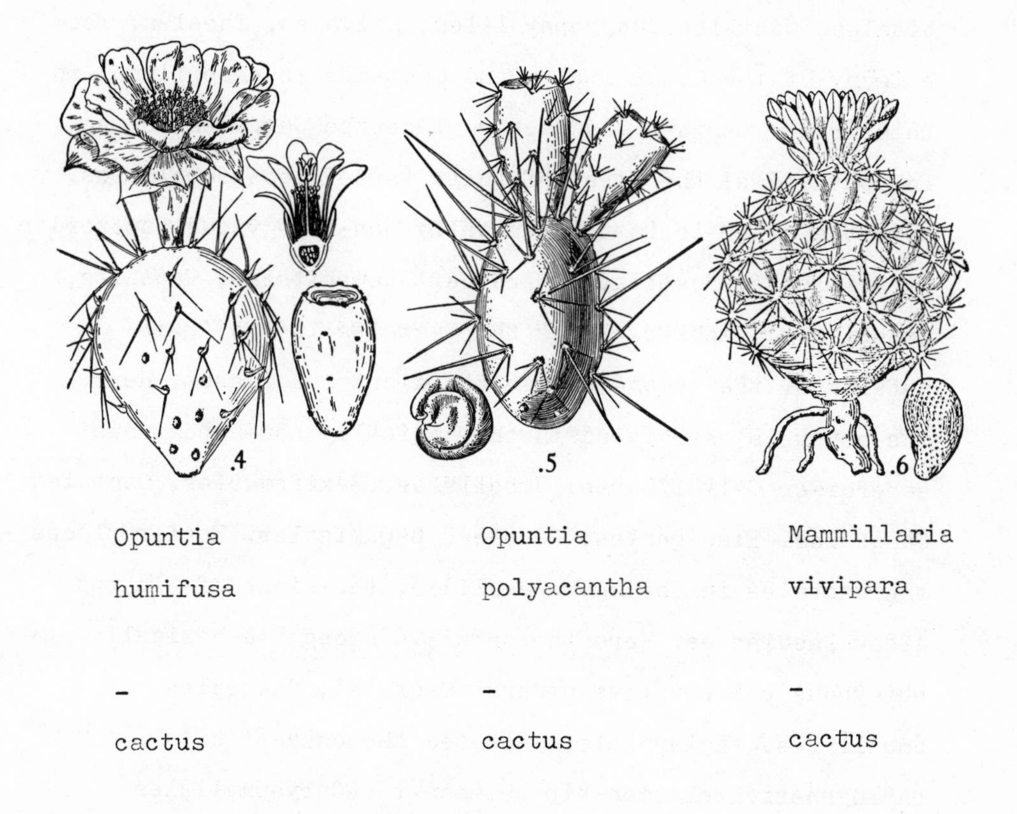

petals. Calyx of numerous sepals. Fruit a berry.
Plants succulent (with spines).

Opuntia sp.

dEpiGi/6AnCnKn/

30 3:1

Endlicher places the Cactales in the Polypetalidae, the third of his three dicot subclasses. Here the order is associated in a thealian order group: Violales, Cucurbitales, Cactales, Caryophyllales, Malvales, Theales. The epigyny of the Cucurbitales and Cactales is exceptional in this group. Bentham and Hooker place the Cactales in the Polypetalidae, the first of their three dicot subclasses. Here the order is in the basically non-hypogynous superorder Calyciflorae: Rosales, Onagrales, Cucurbitales, Cactales, Umbellales. Eichler places the Cactales in the Agamopetalidae, the second of his two dicot subclasses. Here the order is associated in the basically non-hypogynous superorder Calyciflorae: Umbellales, Saxifragales, Cactales, Onagrales, Elaeagnales, Rosales, Leguminales. Engler places the Cactales in the Agamopetalidae, the first of his two dicot subclasses. Here the order is among the basically epigynous polypetalous orders: Cactales, Onagrales, Umbellales. Engler later accepted the concept of a phylogenetic relationship between the Caryophyllales and Cactales, but not to the extent of disturbing his original sequence of orders. Bessey places the Cactales in the Dicotocotyloididae, the second of his two dicot subclasses. Here the order is in the superorder Apopetalae. He diagrams as a phylogenetic sequence: Rosales, Onagrales, Cactales. Hutchinson places the Cactales in the Lignosidae, the first of his two dicot subclasses. Here the order is derived from a line which branches to give rise to both the Cactales and the Cucurbitales. Melchior places the

Cactales in the Agamopetalidae, the first of his two dicot subclasses. Here, however, he follows the logic of Engler's later views in transferring the order from its Englerian position to an association with the Caryophyllales. Cronquist does not recognize the Cactales. Instead he goes a step beyond Melchior's position by placing the Cactaceae within the Caryophyllales.

30 4:1

GLOSSARY FOR CACTALES

Cactales. Order in subclass Agamopetalidae.
  Opuntiae. Classis in cohors Dialypetalae (Endl.)
  Opuntiinae. Unterreihe in Reihe Calyciflorae (Eichl.)
      In Klasse Chori- und Apetalae.
  Opuntiales. Series in subclassis Archichlamydeae (Engl.)
  Cactales. Order in superorder Apopetalae (Bessey).
      In subclass Cotyloideae.
  Cactales. Order in division Lignosae (Hutch.)
  Cactales. Reihe in Unterklasse Archichlamydeae (Melch.)
Cactaceae. Family in order Cactales.
  Cacteae. Ordo in classis Opuntiae (Endl.)
  Cacteae. Ordo in cohors Ficoidales (B.&H.)
  Cacteae. Familie in Unterreihe Opuntiinae (Eichl.)

Cactaceae. Familia in series Opuntiales (Engl.)
Cactaceae. Family in order Cactales (Bessey), (Hutch.)
Cactaceae. Familie in Reihe Cactales (Melch.)
Cactaceae. Family in order Caryophyllales (Cronq.)

ONAGRALES

The Onagrales contain the Elaeagnaceae, Lythraceae, and Onagraceae. The Elaeagnaceae have tetramerous epigynous flowers with unicarpellary gynoecium and uniseriate perianth of tepals united into a tube upon which the stamens are inserted. The indumentum of the foliage is lepidote. The Lythraceae have perigynous flowers with stamens not only arising from within the hypanthium, but being often di- or trimorphic. Petals inserted toward the rim of the hypanthium. The Onagraceae have epigynous tetramerous flowers with elongate hypanthium, both stamens and petals being inserted towards its rim.

Elaeagnus is derived from the ancient Greek name for the plant now known as Salix caprea. It is a compound signifying literally: "olive chaste-tree," alluding to supposed magical properties of the plant suggested by its gray pubescence. Although the plant now designated by the name is unrelated to the ancient one, it fits the name more closely in that the fruit of *Elaeagnus* resembles an olive more than does the fruit of *Salix*, and the silvery indumentum of *Elaeagnus angustifolia*, Russian olive, is more striking than the pubescence of the goat willow. *Lythrum* is derived from an ancient Greek name for the plant now called *Lythrum salicaria*. Literally *Lithrum* signifies: "gore" and is descriptive of an impressionistic aspect of the inflorescence. *Onagra* is an obsolete synonym of *Oenothera* and was adopted from an ancient Greek name for the plant now known as *Nerium oleander*, a shrub of the Apocynaceae. *Oenothera* was adopted from another ancient Greek name. In Greek it is synonymous with *onagra*.

31 2:1

ELAEAGNACEAE

dPeriG1A2-4P/̸4/perf.,unisex.,ach.in fleshyP

Dicotyledonous plants with perigynous flowers. Gynoecium of a single carpel. Androecium of 2 to 4 stamens. Perianth of 4 united tepals. Flowers perfect, or unisexual. Fruit an achene in the fleshy perianth.

Elaeagnus sp.

dPeriG1A4P/̸4/

ELAEAGNACEAE

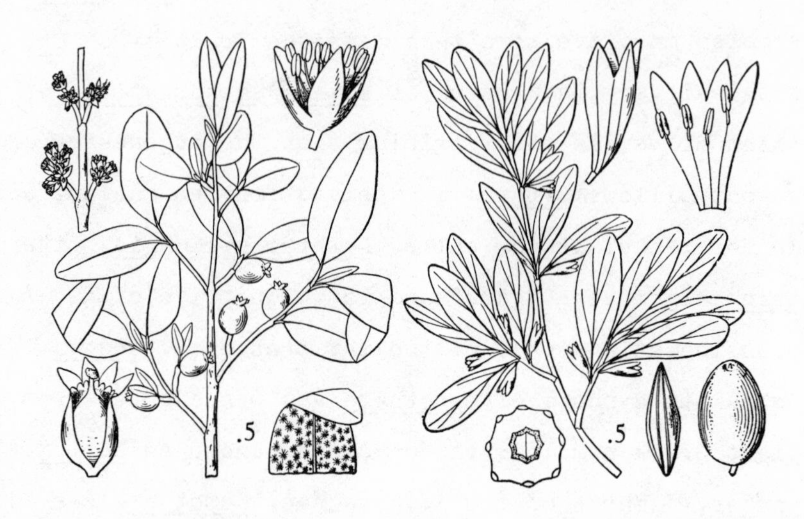

Shepherdia canadensis	Elaeagnus commutata
-	silver berry
buffalo berry	-

LYTHRACEAE

dPeriG⁄2-5A(1-)4-14C4-7,O,K,P,4-7/caps./styles ⁄

Dicotyledonous plants with perigynous flowers. Gynoecium of 2 to 5 united carpels. Androecium of 4 to 14 stamens (or as few as 1). Corolla of 4 to 7 petals, or none. Calyx, or perianth, of 4 to 7 sepals, or tepals. Fruit a capsule. Styles united.

Lythrum sp.

dPeriG1̸2A12C6K6/

LYTHRACEAE

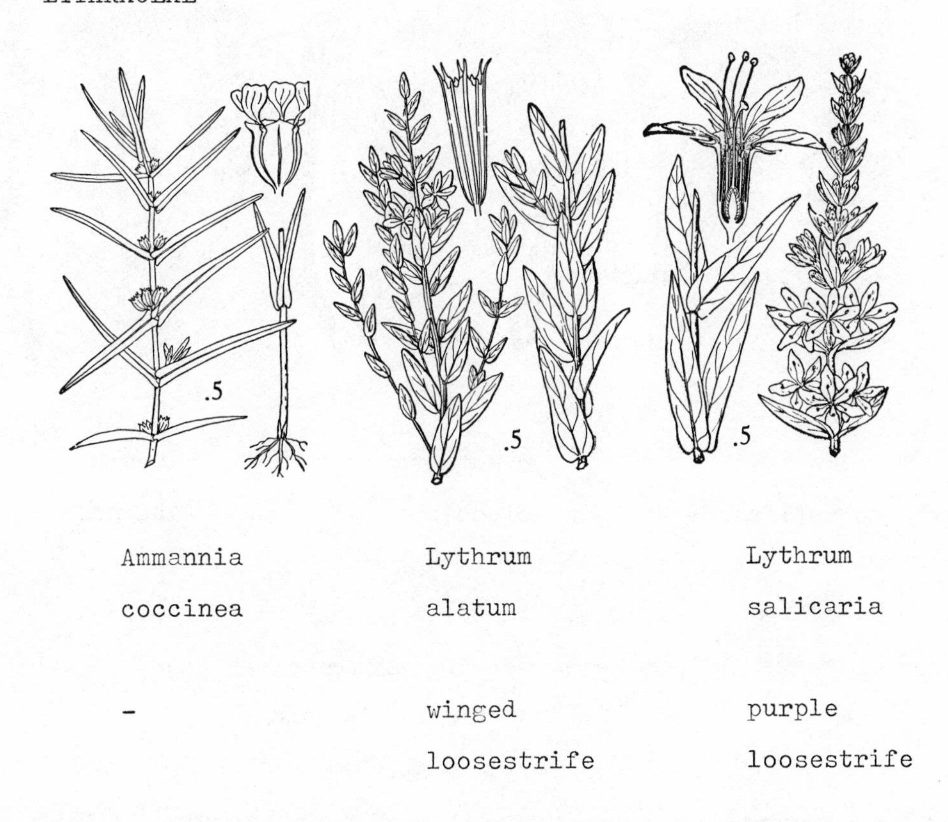

Ammannia coccinea	Lythrum alatum	Lythrum salicaria
-	winged loosestrife	purple loosestrife

ONAGRACEAE

dEpi(oft.w.hypanth.tube)G1̸(2-)4L1-4A(2-)8C(2-)4K1̸,i,

(2-)4/caps./ovules 1-n per L/styles 1̸

Dicotyledonous plants with epigynous flowers (often with

ONAGRACEAE

Epilobium angustifolium	Oenothera biennis	Gaura biennis
-	-	-
	evening primrose	

a hypanthium tube). Gynoecium of 4 united carpels (or as few as 2). Locules 1 to 4. Androecium of 8 stamens (or as few as 2). Corolla of 4 petals (or as few as 2). Calyx of 4 sepals (or as few as 2), united or free. Fruit a capsule. Ovules 1 to many in each locule. Styles united.

Oenothera sp.

dEpi(w.hypanth.tube)Gi̸A8C4K4/

31 3:1

Endlicher places the Onagrales in the Polypetalidae, the third of his three dicot subclasses. Here the order is in the calyciflorous order group: Onagrales, Rosales, Leguminales. Bentham and Hooker place the Onagrales in the Polypetalidae, the first of their three dicot subclasses. Here the order is in the superorder Calyciflorae: Rosales, Onagrales, Cucurbitales, Cactales, Umbellales. Eichler places the Onagrales in the Agamopetalidae, the second of his two dicot subclasses. Here the order is in the superorder Calyciflorae: Umbellales, Saxifragales, Onagrales, Elaeagnales, Rosales, Leguminales. Engler places the Onagrales in the Agamopetalidae, the first of his two dicot subclasses. Here the order is among the basically epigynous polypetalous orders: Cactales, Onagrales, Umbellales. Bessey places the Onagrales in the Dicotocoyloididae, the second of his two dicot subclasses. Here the order is in the superorder Apopetalidae. Bessey diagrams as a phylogenetic sequence: Rosales, Onagrales. Hutchinson places the Onagrales in the Herbacidae, the second of his two dicot subclasses. He diagrams as a phylogenetic sequence: Caryophyllales, Onagrales. Melchior maintains the Onagrales in their

Englerian position. Here the order retains its association with the Umbellales. Cronquist places the Onagrales in the Rosidae, the fifth of his six dicot subclasses. He diagrams as a phylogenetic sequence: Rosales, Onagrales.

Endlicher recognizes the Onagrales as containing families in the sequence: Onagraceae, Lythraceae. The Elaeagnaceae he assigns to the Elaeagnales, an order of his subclass Gamopetalidae. This order contains also the Santalaceae. On this interpretation the Elaeagnaceae are assumed to lack a hypanthium. Bentham and Hooker prefer the sequence: Lythraceae, Onagraceae. They assign the Elaeagnaceae to the Elaeagnales, an order of the subclass Apetalidae. Eichler follows Endlicher's sequence: Onagraceae, Lythraceae. He assigns the Elaeagnaceae to the Elaeagnales, another order of the subclass Agamopetalidae and directly following the Onagrales. Engler recognizes the Onagrales as containing families in the sequence: Elaeagnaceae, Lythraceae, Onagraceae. Bessey, Melchior, and Cronquist follow the Bentham and Hooker sequence: Lythraceae, Onagraceae. Bessey assigns the Elaeagnaceae to the Celastrales, another order of the superorder Apopetalidae within the subclass Dicotocotyloididae. Bessey diagrams the Onagrales and Celastrales as two lines separately arising from the Rosales. Hutchinson assigns the Elaeagnaceae and Lythraceae to the Rhamnales and Myrtales of his Lignosidae. Melchior assigns the Elaeagnaceae to the Elaeagnales, another order of the subclass Agamopetalidae. Here the order is interpreted as lacking a hypanthium and therefore as being hypogynous with epipetalous stamens. The group is therefore placed near the Malvales. Cronquist assigns the Elaeagnaceae to the Elaeagnales, another order of the subclass Rosidae. He diagrams as a phylogenetic sequence: Rosales, Onagrales, Elaeagnales.

31 4:1

GLOSSARY FOR ONAGRALES

Elaeagnales. Order in subclass Agamopetalidae.
  Thymeleae. Classis in cohors Apetalae (Endl.)
  Daphnales. Series in subclassis Monochlamydeae (B.&H.)
  Thymelaeinae. Unterreihe in Reihe Calyciflorae (Eichl.)
    In Klasse Chori- und Apetalae.
  Thymelaeales. Reihe in Unterklasse Archychlamydeae (Melch.)
  Proteales. Order in subclass Rosidae (Cronq.)
Onagrales. Order in subclass Agamopetalidae.
  Calyciflorae. Classis in cohors Dialypetalae (Endl.)
  Myrtales. Cohors in series Calyciflorae (B.&H.)
    In subclassis Polypetalae.

Myrtiflorae. Unterreihe in Reihe Calyciflorae (Eichl.)
In Klasse Chori- und Apetalae.
Myrtiflorae. Series in subclassis Archichlamydeae (Engl.)
Myrtales. Order in superorder Apopetalae (Bessey).
In subclass Cotyloideae.
Onagrales. Order in division Herbaceae (Hutch.)
Myrtiflorae. Reihe in Unterklasse Archichlamydeae (Melch.)
Myrtales. Order in subclass Rosidae (Cronq.)

Nyctaginales. Order in subclass Agamopetalidae.
Thymelaeales. Order in division Lignosae (Hutch.)

Elaeagnaceae. Family in order Elaeagnales.
Elaeagneae. Ordo in classis Thymeleae (Endl.)
Elaeagnaceae. Ordo in series Daphnales (B.&H.)
Elaeagnaceae. Familie in Unterreihe Thymelaeinae (Eichl.)
Elaeagnaceae. Familia in series Myrtiflorae (Engl.)
Elaeagnaceae. Family in order Celastrales (Bessey).
Elaeagnaceae. Family in order Rhamnales (Hutch.)
Elaeagnaceae. Familie in Reihe Thymelaeales (Melch.)
Elaeagnaceae. Family in order Proteales (Cronq.)

Lythraceae. Family in order Onagrales.
Lythrarieae. Ordo in classis Calyciflorae (Endl.)
Lythrarieae. Ordo in cohors Myrtales (B.&H.)
Lythraceae. Familie in Unterreihe Myrtiflorae (Eichl.)
Lythraceae. Familia in series Myrtales (Engl.)
Lythraceae. Family in order Myrtales (Bessey), (Hutch.), (Cronq.)

Lythraceae. Familie in Reihe Myrtiflorae (Melch.)

Onagraceae. Family in order Onagrales.

Oenothereae. Ordo in classis Calyciflorae (Endl.)

Onagrarieae. Ordo in cohors Myrtales (B.&H.)

Onagraceae. Familie in Unterreihe Myrtiflorae (Eichl.)

Onagraceae. Familia in series Myrtales (Engl.)

Oenotheraceae. Family in order Myrtales (Bessey).

Onagraceae. Family in order Onagrales (Hutch.)

Onagraceae. Familie in Reihe Myrtiflorae (Melch.)

Onagraceae. Family in order Myrtales (Cronq.)

Santalaceae. See 16 4.

Santalaceae. Ordo in classis Thymeleae (Endl.)

Aristolochiaceae. See 17 4.

Aristolochiaceae. Family in order Myrtales (Bessey).

Nyctaginaceae. See 19 4.

Nyctaginaceae. Family in order Thymelaeales (Hutch.)

32 1:1

UMBELLALES

The Umbellales include the Araliaceae, Umbelliferae, and Cornaceae. The Araliaceae have epigynous flowers in umbellate inflorescences. The Umbelliferae have epigynous flowers in compound umbellate inflorescences. The Cornaceae have clustered tetramerous epigynous flowers. In some species of *Cornus* tetramerous white bracts transform the clusters into pseudanthia.

The name *Aralia* is a Latinization of the Canadian French *aralie* used for specimens sent from Canada to Tournefort. An Iroquoian origin for the name has been suggested. Umbelliferae is a descriptive family name based on the modern pre-Linnaean Latin adjective *umbelliferus*, which signifies: "plants bearing umbels." The ancient Latin *umbella* signifies "umbrella" and is derived as a diminutive from *umbra*, "shade." The Latin element -*fer*- is cognate with English *bear*. Note the correspondence of letters: f/b, e/ea, r/r. Linnaeus used the name *Umbellatae* for this group. *Cornus* signifies "horn" or "horny wood," in allusion to the use of the wood of the plants of this group for skewers. The English name *dogwood* is explained as *dag*, "skewer," and *wood*. The element *dag*- is unrelated to the word *dog*, but it occurs in the word *dagger*.

32 2:1

ARALIACEAE

dEpiG/2-5A5C,P,i,_,5K5,O/polygam.,dioec.;ber.,drup./ovules 1 per L/styles±i

Dicotyledonous plants with epigynous flowers. Gynoecium of 2 to 5 united carpels. Androecium of 5 stamens.

ARALIACEAE

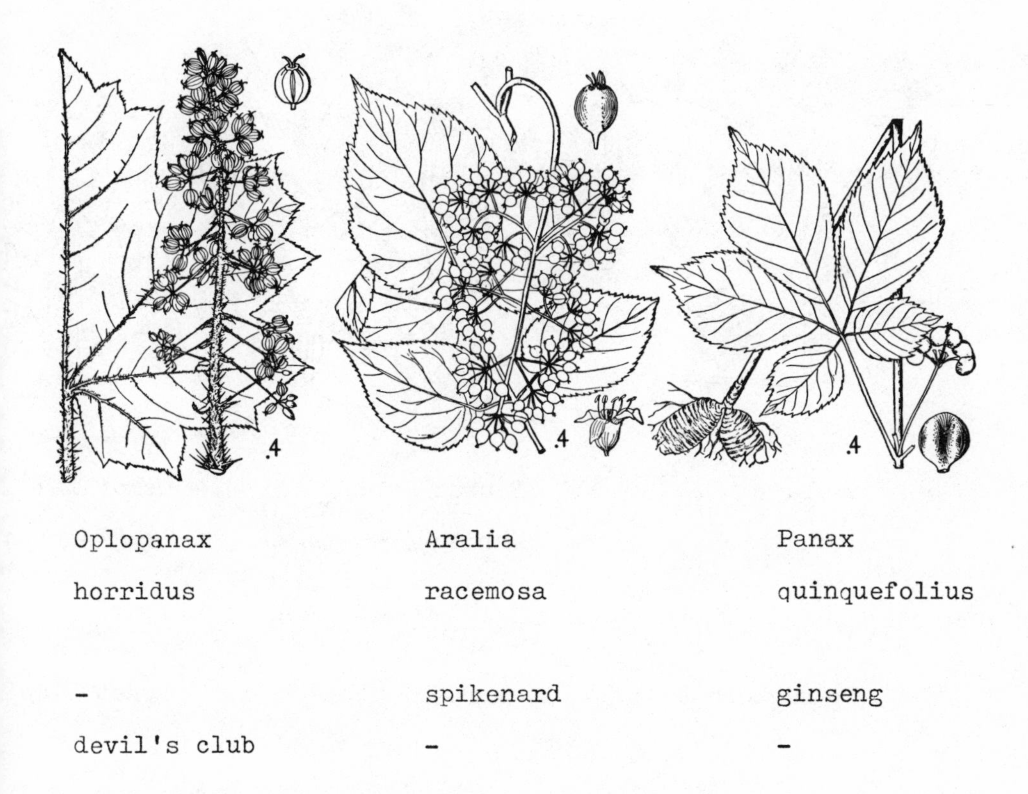

Oplopanax	Aralia	Panax
horridus	racemosa	quinquefolius
-	spikenard	ginseng
devil's club	-	-

Corolla, or perianth, of 5 petals, or tepals, free or united at the base. Calyx of 5 sepals, or none. Plants polygamous, or dioecious. Fruit a berry, or drupaceous. Ovules 1 in each locule. Styles more or less free.

Aralia sp.

dEpiG/5A5C5K5/

UMBELLIFERAE

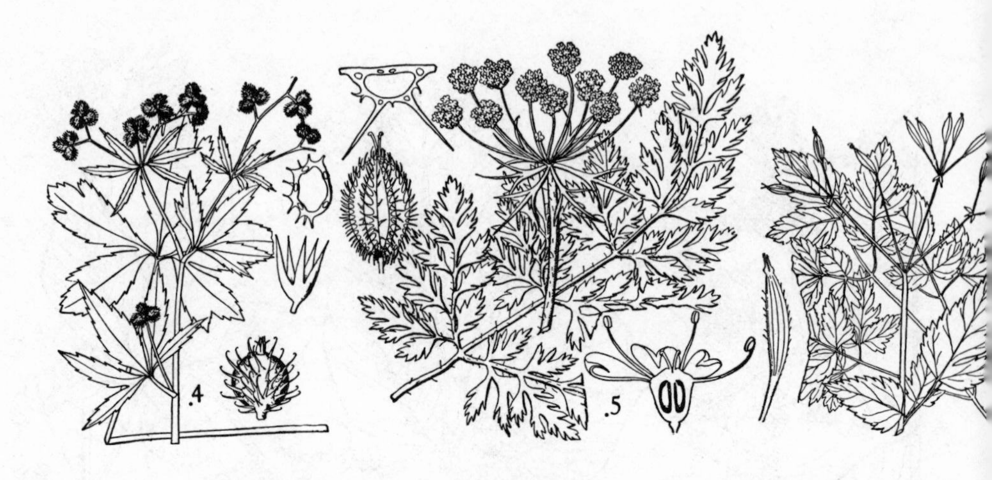

Sanicula canadensis	Daucus carota	Osmorhiza longistylis
-	-	-
black snakeroot	carrot	sweet cicel

UMBELLIFERAE

dEpi(w.disk)G/2A5C,P,5K5,0/schizoc./ovules 1 per L

Dicotyledonous plants with epigynous flowers (having a disk). Gynoecium of 2 united carpels. Androecium of 5 stamens. Corolla, or perianth, of 5 petals, or tepals. Calyx of 5 sepals, or none. Fruit a schizocarp. Ovules 1 in each locule.

Daucus carota

dEpi(w.disk)G/2A5C5K5/

CORNACEAE

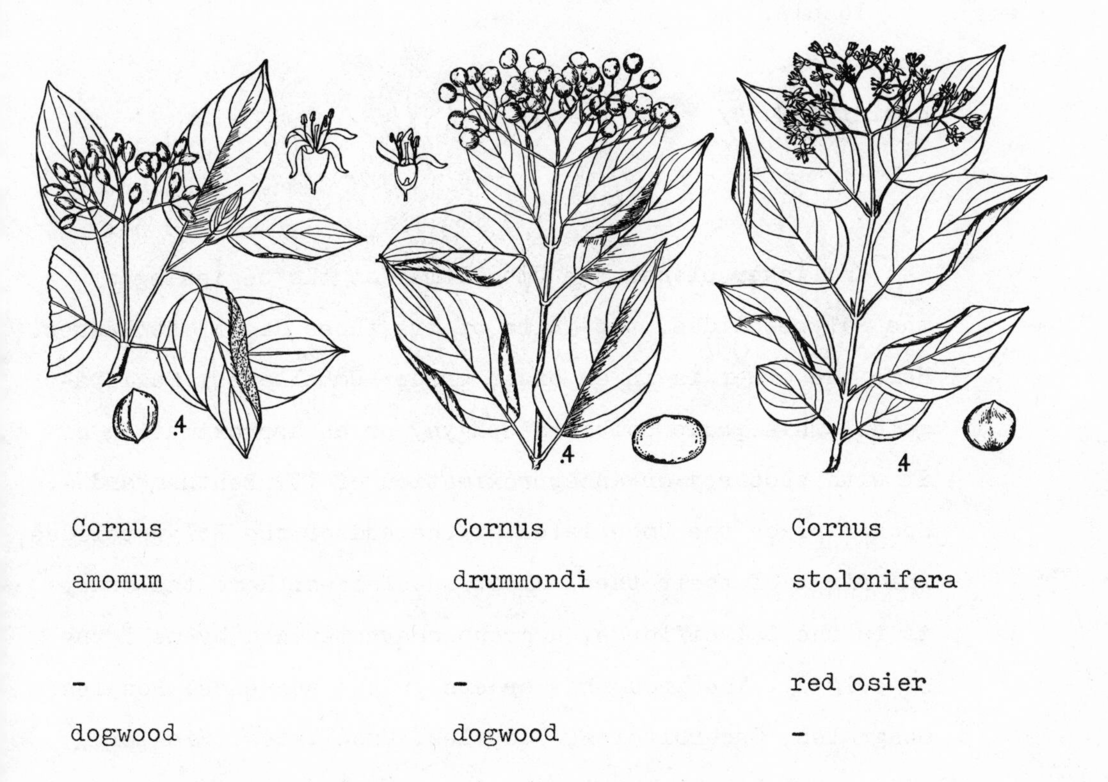

Cornus	Cornus	Cornus
amomum	drummondi	stolonifera
-	-	red osier
dogwood	dogwood	-

CORNACEAE

dEpiG/2,1,A4-12C4-5,0,K,P,4-5/perf.,dioec.,drup./
ovules 1 per L

Dicotyledonous plants with epigynous flowers. Gynoecium of 2 united carpels, or of a single carpel. Androecium of 4 to 12 stamens. Corolla of 4 to 5 petals, or none. Calyx, or perianth, of 4 to 5 sepals, or tepals. Flowers perfect, or plants dioecious. Fruit drupaceous. Ovules 1 in each locule.

Cornus sp.

dEpiG/2A4C4K4/

32 3:1

Endlicher places the Umbellales at the beginning of the Polypetalidae, the third of his three dicot subclasses. Here the order is in an order group: Umbellales, Saxifragales. This group combines epigyny or an approximation of it with apocarpy or an approximation of it. Bentham and Hooker place the Umbellales at the end of the Polypetalidae, the first of their three dicot subclasses. Here the order is in the Calyciflorae, a group characterized by perigyny or epigyny. The group has orders in the sequence: Rosales, Onagrales, Cucurbitales, Cactales, Umbellales. Eichler places the Umbellales in the Agamopetalidae, the second of his two dicot subclasses. Here the order is at the beginning of the superorder Calyciflorae. This runs: Umbellales, Saxifragales, Onagrales, Elaeagnales, Rosales, Leguminales. Note that this sequence incorporates Endlicher's order group: Umbellales, Saxifragales. Engler places the Umbellales at the end of the Agamopetalidae, the first of his two dicot subclasses. Here the order is associated with the basically epigynous polypetalous orders: Cactales, Onagrales, Umbellales. Bessey places the Umbellales in the

Dicotocotyloididae, the second of his two dicot subclasses. Here the order stands at the end of the superorder Apopetalae. He diagrams as a phylogenetic sequence: Magnoliales, Rosales, Celastrales, Umbellales. Hutchinson splits the umbellalian taxa of other systems into woody and herbaceous groups. The woody group he places in the order Cornales within his Lignosidae, the first of his two dicot subclasses. This order contains not only the Cornaceae and Araliaceae but also the Caprifoliaceae. He diagrams the Cornales as indirectly derived from the Rosales. The herbaceous group he places in a narrowly circumscribed order Umbellales within the subclass Herbacidae. This order contains only the Umbellaceae. He diagrams as a phylogenetic sequence: Ranunculales, Saxifragales, Umbellales. Melchior follows the Englerian system in his placement of the Umbellales. Cronquist, like Hutchinson, splits the umbellalian taxa of most systems into the two orders Cornales and Umbellales. He places both of these orders within the subclass Rosidae, the fifth of his six dicot subclasses. Here the orders stand remote from each other. The orders themselves differ from Hutchinson's in that the family Araliaceae falls in the Umbellales instead of the Cornales. Cronquist diagrams as separate phylogenetic sequences: Rosales, Sapindales, Umbellales and Rosales, Cornales.

Endlicher arranges the families of the Umbellales in the sequence: Umbelliferae, Araliaceae, Vitaceae, Cornaceae,

Hamamelidaceae. Bentham and Hooker and Eichler have the sequence: Umbelliferae, Araliaceae, Cornaceae. Engler and Bessey put the Araliaceae first, giving the sequence: Araliaceae, Umbelliferae, Cornaceae. Melchior affirms the Araliaceae to be the most primitive family of the order.

Six of the systems agree on an affinity among the Araliaceae, Umbelliferae, and Cornaceae. Hutchinson dissents from this view by excluding the Umbelliferae. Cronquist dissents by excluding the Cornaceae.

32 4:1

GLOSSARY FOR UMBELLALES

Umbellales. Order in subclass Agamopetalidae.

Discanthae. Classis in cohors Dialypetalae (Endl.)

Umbellales. Cohors in series Calyciflorae (B.&H.)

In subclassis Polypetalae.

Umbelliflorae. Unterreihe in Reihe Calyciflorae (Eichl.)

In Klasse Chori- und Apetalae.

Umbelliflorae. Series in subclassis Archichlamydeae (Engl.)

Umbellales. Order in superorder Apopetalae (Bessey).

In subclass Cotyloideae.

Umbellales. Order in division Herbaceae (Hutch.)

Umbelliflorae. Reihe in Unterklasse Archichlamydeae (Melch.)

Umbellales. Order in subclass Rosidae (Cronq.)

Cornales. Order in subclass Agamopetalidae.

Araliales. Order in division Lignosae (Hutch.)

Cornales. Order in subclass Rosidae (Cronq.)

Araliaceae. Family in order Umbellales.

Araliaceae. Ordo in classis Discanthae (Endl.)

Araliaceae. Ordo in cohors Umbellales (B.&H.)

Araliaceae. Familie in Unterreihe Umbelliflorae (Eichl.)

Araliaceae. Familia in series Umbelliflorae (Engl.)

Araliaceae. Family in order Umbellales (Bessey), (Cronq.)

Araliaceae. Family in order Araliales (Hutch.)

Araliaceae. Familie in Reihe Umbelliflorae (Melch.)

Umbelliferae. Family in order Umbellales.

Umbelliferae. Ordo in classis Discanthae (Endl.)

Umbelliferae. Ordo in cohors Umbellales (B.&H.)

Umbelliferae. Familie in Unterreihe Umbelliflorae (Eichl.)

Umbelliferae. Familia in series Umbelliflorae (Engl.)

Apiaceae. Family in order Umbellales (Bessey), (Hutch.)

Umbelliferae. Familie in Reihe Umbelliflorae (Melch.)

Umbelliferae. Family in order Umbellales (Cronq.)

Cornaceae. Family in order Umbellales.

Corneae. Ordo in classis Discanthae (Endl.)

Cornaceae. Ordo in cohors Umbellales (B.&H.)

Cornaceae. Familie in Unterreihe Umbelliflorae (Eichl.)

Cornaceae. Familia in series Umbelliflorae (Engl.)

Cornaceae. Family in order Umbellales (Bessey).

Cornaceae. Family in order Araliales (Hutch.)

Cornaceae. Familie in Reihe Umbelliflorae (Melch.)

Cornaceae. Family in order Cornales (Cronq.)

Hamamelidaceae. See 24 4.

Hamamelideae. Ordo in classis Discanthae (Endl.)

Vitaceae. See 27 4.

Ampelideae. Ordo in classis Discanthae (Endl.)

Caprifoliaceae. See 42 4.

Caprifoliaceae. Family in order Araliales (Hutch.)

PENTACYCLIC GAMOPETALOUS ORDER GROUPS

The orders of Endlicher's subclass Gamopetalidae run as follows: Plumbaginales, Asterales, Campanulales, Rubiales, Gentianales, Labiatales, Polemoniales, Scrophulariales, Primulales, Ericales. Here the orders Primulales (incl. Ebenaceae) and Ericales constitute a pentacyclic order group. The orders of the Bentham and Hooker subclass Gamopetalidae are distributed into the superorders: Inferae, Heteromerae, and Bicarpellatae. The superorder Heteromerae is pentacyclic and contains the orders: Ericales, Primulales, Ebenales. The orders of Eichler's subclass Gamopetalidae are distributed into the superorders: Haplostemones, Diplostemones, and Obdiplostemones. The superorders Diplostemones and Obdiplostemones are pentacyclic. The Diplostemones contain the orders Primulales and Ebenales. The Obdiplostemones contain the order Ericales. The orders of Engler's subclass Gamopetalidae are arranged in two principal aggregations. The first of these is pentacyclic and includes two order groups. One group has flowers with two whorls of stamens or a single whorl opposite the petals and includes the Ericales and Primulales. The other group has flowers usually with numerous stamens and consists of the Ebenales. The orders of Bessey's subclass

Dicotostrobiloididae are distributed into the superorders: Apopetalae-Polycarpellatae, Sympetalae-Polycarpellatae, and Sympetalae-Dicarpellatae. The Sympetalae-Polycarpellatae constitute a pentacyclic group which contains the orders: Ebenales, Ericales, Primulales. Hutchinson's dicot subclass Lignosidae contains no pentacyclic gamopetalous order group. Instead within this subclass the Ericales are treated as a climax order with derivation from the Theales, and the Ebenales are treated as a climax order with derivation from the Rhamnales. Within the subclass Herbacidae the Primulales are treated as derived from the Gentianales. The orders of Melchior's subclass Gamopetalidae are arranged in two principal aggregations. The first of these is basically pentacyclic and includes the orders: Ericales, Primulales, and Ebenales. Melchior views the group as based on morphological similarities, but he does not recognize phylogenetic relationships between its members. The orders of Cronquist's subclass Theidae are arranged in two principal order groups. The second of these is pentacyclic gamopetalous and includes the orders: Ericales, Ebenales, and Primulales.

GLOSSARY FOR PENTACYCLIC GAMOPETALOUS SUPERORDERS, ETC.

Heteromerae (cohortes: Ericales, Primulales, Ebenales).
Series in subclassis Gamopetalae (B.&H.)

Diplostemones (Unterreihen: Primulinae, Diospyrinae).
Reihe in Klasse Sympetalae (Eichl.)

Obdiplostemones (Unterreihe: Bicornes (Ericaceae, etc.))
Reihe in Klasse Sympetalae (Eichl.)

Sympetalae-Polycarpellatae (orders: Ebenales, Ericales, Primulales). Superorder in subclass Strobiloideae (Bessey).

33 1:1

ERICALES

The only family of the Ericales here considered is the Ericaceae. The order is commonly classified as being gamopetalous and pentacyclic, but it does not strictly conform to these characteristics. Some of its genera have petals distinct, and some are tetracyclic with obdiplostemony—the position of the stamens opposite the petals, implying a lost whorl of alternate stamens.

The name Erica is derived through ancient Latin from a Greek name for the plant now known as Erica arborea.

33 2:1

ERICACEAE

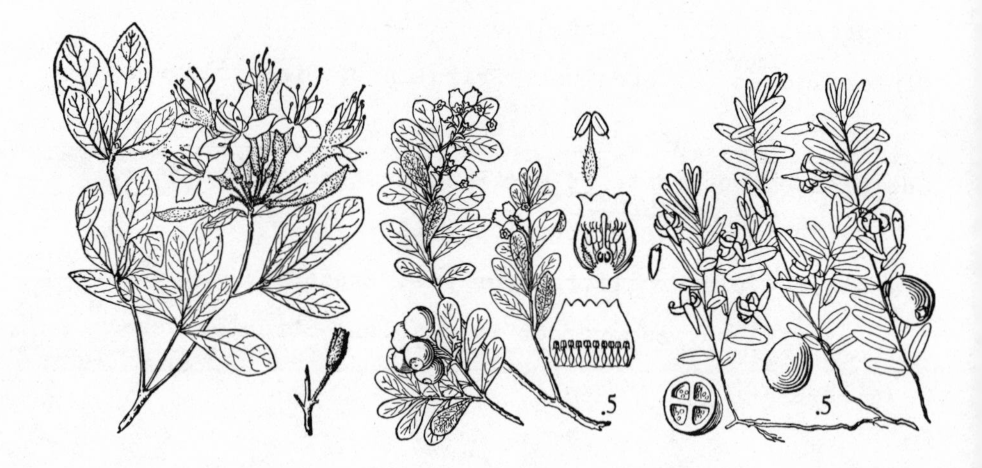

Rhododendron viscosum	Arctostaphylos uva-ursi	Vaccinium macrocarpon
swamp azalea	- bearberry	cranberry -

ERICACEAE

dHypo,Epi,G/2-5L2-5,10,A4-10(4-5(opp.),(4-5)+(4-5(obdipl.)))

C/4-5K_,i,4-5/caps.,ber.,drup.

Dicotyledonous plants with hypogynous, or epigynous, flowers.

Gynoecium of 2 to 5 united carpels and 2 to 5 locules, or 10. Androecium of 4 to 10 stamens (4 to 5 stamens (opposite the petals), or of both an inner group of 4 to 5 stamens and an equal outer group (obdiplostemonous)). Corolla of 4 to 5 united petals. Calyx of 4 to 5 sepals, united at the base or free. Fruit a capsule, berry, or drupaceous.

Erica sp.

dHypoG/4A8(4+4(obdipl.))C/4K4/

33 3:1

Endlicher places the Ericales at the end of the Gamopetalidae, the second of his three dicot subclasses. Here the order is in a pentacyclic order group: Primulales (incl. Ebenaceae), Ericales. Bentham and Hooker place the Ericales in the Gamopetalidae, the second of their three dicot subclasses. Here the order is in the superorder Heteromerae: Ericales, Primulales, Ebenales. The name of the superorder alludes to the variability of carpel number within the group. Eichler places the Ericales at the end of the Gamopetalidae, the first of his two dicot subclasses. He classifies the order as the only member here considered of the superorder Obdiplostemones. Engler places the Ericales in the Gamopetalidae, the second of his two dicot subclasses. The order is in a pentacyclic order group: Ericales, Primulales. Bessey places the Ericales in the

Dicotostrobiloididae, the first of his two dicot subclasses. Here the order is in the superorder Sympetalae-Polycarpellatae: Ebenales, Ericales, Primulales. He diagrams as a phylogenetic sequence: Magnoliales, Caryophyllales, Ericales. Hutchinson places the Ericales in the Lignosidae, the first of his two dicot subclasses. Here the order stands remote from the Primulales and even far from the Ebenales. He diagrams as a phylogenetic sequence: Magnoliales, Theales, Ericales. Melchior classifies the Ericales in the Gamopetalidae, the second of his two dicot subclasses. Here they stand in a group of orders with morphological similarities: Ericales, Primulales, and Ebenales. Melchior does not, however, recognize phylogenetic relationships between any of these orders. Cronquist places the Ericales in the Theidae, the fourth of his six dicot subclasses. He diagrams a phylogenetic line from the Theales to the Ericales.

Hutchinson and Cronquist agree in deriving the Ericales from the Theales. Hutchinson emphasizes hypogynous stamens in gamopetalous flowers as a primitive feature of the Ericaceae.

GLOSSARY FOR ERICALES

Ericales. Order in subclass Gamopetalidae.

Bicornes. Classis in cohors Gamopetalae (Endl.)

Ericales. Cohors in series Heteromerae (B.&H.)

In subclassis Gamopetalae.

Bicornes. Unterreihe in Reihe Obdiplostemones (Eichl.)

In Klasse Sympetalae.

Ericales. Series in subclassis Metachlamydeae (Engl.)

Ericales. Order in superorder Sympetalae-Polycarpellatae

(Bessey). In subclass Strobiloideae.

Ericales. Order in division Lignosae (Hutch.)

Ericales. Reihe in Unterklasse Sympetalae (Melch.)

Ericales. Order in subclass Dilleniidae (Cronq.)

Ericaceae. Family in order Ericales.

Ericaceae. Ordo in classis Bicornes (Endl.)

Ericaceae. Ordo in cohors Ericales (B.&H.)

Ericaceae. Familie in Unterreihe Bicornes (Eichl.)

Ericaceae. Familia in series Ericales (Engl.)

Ericaceae. Family in order Ericales (Bessey), (Hutch.),

(Cronq.)

Ericaceae. Familie in Reihe Ericales (Melch.)

Diapensiaceae. See 34 4.

Diapensiaceae. Ordo in cohors Ericales (B.&H.)

Diapensiaceae. Familie in Unterreihe Bicornes (Eichl.)
Diapensiaceae. Familia in series Ericales (Engl.)
Diapensiaceae. Family in order Ericales (Bessey), (Hutch.)

34 1:1

DIAPENSIALES

The Diapensiales are restricted to the single family Diapensiaceae. The family is inconspicuous in our area, and the group is not given regular coverage in the present survey. The Diapensiaceae are commonly classified as pentacyclic and gamopetalous, but neither of these characters is always obvious. The inner staminal whorl is commonly staminodial, rudimentary, or absent; and the corolla is in some forms cleft nearly to the base.

Diapensia is a name composed of Greek elements and apparently signifying: "plant having flowers with parts in fives." The applicability of this signification is obscure.

34 2:1

DIAPENSIACEAE

dHypoG%3Ai5(epipet.),(SA)%10(epipet.)(S5+A5)C%,_,5
K%,_,5/caps.

DIAPENSIACEAE

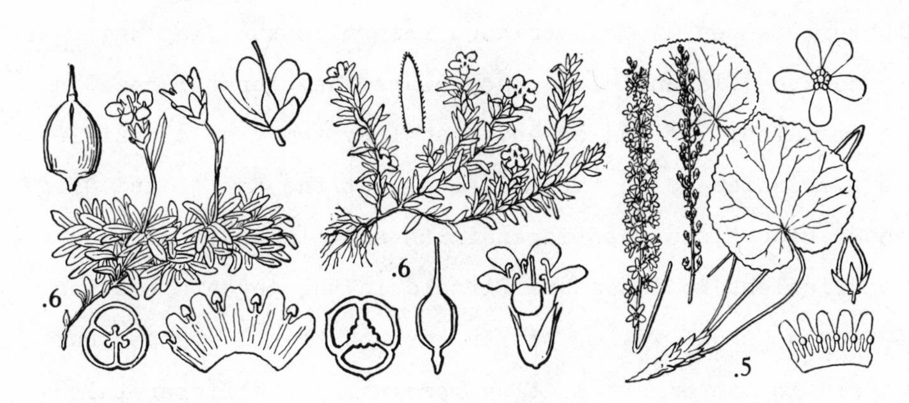

Diapensia lapponica	Pyxidanthera barbulata	Galax aphylla
-	-	-

Dicotyledonous plants with hypogynous flowers. Gynoecium of 3 united carpels. Androecium of 5 distinct stamens (epipetalous), or of a total of 10 united staminodia and stamens (epipetalous), (a cycle of 5 staminodia and a cycle of 5 stamens). Corolla of 5 petals, united or joined at the base. Calyx of 5 sepals, united or joined at the base. Fruit a capsule.

Galax aphylla

dHypoG/3(SA)/10(epipet.)(S5+A5)C_5K_5/

Endlicher does not recognize the Diapensiaceae but places <u>Diapensia</u> in the Ericaceae. Bentham and Hooker, Eichler, Bessey, and Hutchinson recognize the Diapensiaceae, but not the Diapensiales. They place the Diapensiaceae in the Ericales. The Engler and Prantl system also classifies the Diapensiaceae in the Ericales, but the Engler and Gilg system recognizes the Diapensiales and places them ahead of the Ericales. Robinson and Fernald in the seventh edition of Gray's <u>Manual</u> follow the Engler and Prantl system, but Fernald in the eighth edition recognizes the Diapensiales while retaining the Diapensiaceae in their Engler and Prantl position. Melchior follows the Engler and Gilg system in placing the Diapensiales ahead of the Ericales. Cronquist places the Diapensiales after the Ericales in the Theidae, the fourth of his six dicot subclasses. He diagrams the Diapensiales as an offshoot on a phylogenetic line from the Theales to the Ericales.

GLOSSARY FOR DIAPENSIALES

Diapensiales. Order in subclass Gamopetalidae.

Diapensiales. Reihe in Unterklasse Sympetalae (Melch.)

Diapensiales. Order in subclass Dilleniidae (Cronq.)

Diapensiaceae. Family in order Diapensiales.

Diapensiaceae. Ordo in cohors Ericales (B.&H.)

Diapensiaceae. Familie in Unterreihe Bicornes (Eichl.)

Diapensiaceae. Familia in series Ericales (Engl.)

Diapensiaceae. Family in order Ericales (Bessey), (Hutch.)

Diapensiaceae. Familie in Reihe Diapensiales (Melch.)

Diapensiaceae. Family in order Diapensiales (Cronq.)

35 1:1

PRIMULALES

The Primulales are an order of herbaceous dicotyledons with only one family, the Primulaceae, represented in our area. This family has a gynoecium of typically five united carpels. It has a single locule and a single style and stigma. The placentation is free-central. The fruit is a capsule with apical valves often indicative of carpellary number. The stamens are five and epipetalous, opposite the lobes of the corolla. Their position is taken as evidence that an outer whorl of stamens has been lost. The calyx is synsepalous and five-lobed.

Primula is a medieval Latin substantive based on the nominative singular feminine of the ancient Latin adjective *primulus*, "first," itself a diminutive of primus, "first."

Primula signifies etymologically: "first plant to flower in spring," and the name was applied to the cowslip now having the binomial *P. veris*. This binomial was adopted by Linnaeus from usage as early as 1101, a time when the name *Primula veris* was mentioned in connection with the use of the cowslip in the treatment of paralysis. The epithet *veris* is the genitive of Latin *ver*, "spring."

35 2:1

PRIMULACEAE

Primula mistassinica	Lysimachia nummularia	Anagallis arvensis
-	moneywort	-
primrose	-	pimpernel

PRIMULACEAE

dHypoG⁄5,q,L1A5(epipet.opp.)C⁄5K⁄5/caps.plac.free-centr., bas.

Dicotyledonous plants with hypogynous flowers. Gynoecium of 5 united carpels, or the number of carpels not evident. Locule single. Androecium of 5 stamens (epipetalous and opposite the petals). Corolla of 5 united petals. Calyx of 5 united sepals. Fruit a capsule. Placentation free-central, or basal.

Primula sp.

dHypoG⁄5L1A5(epipet.opp.)C⁄5K⁄5/

35 3:1

Endlicher places the Primulales in the Gamopetalidae, the second of his three dicot subclasses. Here the order is in a pentacyclic order group: Primulales (incl. Ebenaceae), Ericales. Bentham and Hooker place the Primulales in the Gamopetalidae, the second of their three dicot subclasses. Here the order is in the superorder Heteromerae: Ericales, Primulales, Ebenales. The name of the superorder alludes to the variability of carpel number within the group. Eichler places the Primulales in the Gamopetalidae, the first of his two dicot subclasses. Here the order is in the superorder Diplostemones: Primulales, Ebenales. Engler places the Ericales in the Gamopetalidae, the second of his two dicot subclasses. Here the order is in a pentacyclic order group: Ericales, Primulales. Bessey places the Primulales in the Dicotostrobiloididae, the first of his two dicot subclasses. Here the order is in the superorder Sympetalae-Polycarpellatae: Ebenales, Ericales, Primulales. He diagrams as a phylogenetic sequence: Magnoliales, Caryophyllales, Primulales. Hutchinson places the Primulales in the Herbacidae, the second of his two dicot subclasses. He diagrams as a phylogenetic sequence: Caryophyllales, Gentianales, Primulales. The importance of stamen position opposite the lobes of the corolla is thus minimized by Hutchinson. It is to be noted that the Primulales in the subclass Herbacidae stand remote from the Ericales and Ebenales which are in the subclass Lignosidae. Melchior classifies the Primulales in the Gamopetalidae, the second of his two dicot subclasses. Here they stand in a group of orders with morphological similarities: Ericales, Primulales, and Ebenales. Melchior does not, however, recognize phylogenetic relationships between any of these orders. Cronquist places the Primulales in the Theidae, the fourth of his six dicot subclasses. He diagrams a phylogenetic line leading from the Theales and branching to give rise to both the Ebenales and the Primulales.

Endlicher, as stated above, includes the Ebenaceae in the Primulales. In several systems the Plumbaginaceae, as specified under 36 3, are included in the Primulales. In one system the Plantaginaceae, as specified under 41 3, are included in the Primulales.

All of the systems except that of Hutchinson place the Primulales fairly close to the Ericales and Ebenales.

35 4:1

GLOSSARY FOR PRIMULALES

Primulales. Order in subclass Gamopetalidae.

Petalanthae. Classis in cohors Gamopetalae (Endl.)

Primulales. Cohors in series Heteromerae (B.&H.) In subclassis Gamopetalae.

Primulinae. Unterreihe in Reihe Diplostemones (Eichl.) In Klasse Sympetalae.

Primulales. Series in subclassis Metachlamydeae (Engl.)

Primulales. Order in superorder Sympetalae-Polycarpellatae (Bessey). In subclass Strobiloideae.

Primulales. Order in division Herbaceae (Hutch.)

Primulales. Reihe in Unterklasse Sympetalae (Melch.)

Primulales. Order in subclass Dilleniidae (Cronq.)

Primulaceae. Family in order Primulales.

Primulaceae. Ordo in classis Petalanthae (Endl.)

Primulaceae. Ordo in cohors Primulales (B.&H.)

Primulaceae. Familie in Unterreihe Primulinae (Eichl.)

Primulaceae. Familia in series Primulales (Engl.)

Primulaceae. Family in order Primulales (Bessey), (Hutch.), (Cronq.)

Primulaceae. Familie in Reihe Primulales (Melch.)

Plumbaginaceae. See 36 4.

Plumbaginaceae. Ordo in cohors Primulales (B.&H.)

Plumbaginaceae. Familie in Unterreihe Primulinae (Eichl.)

Plumbaginaceae. Familia in series Primulales (Engl.)

Plumbaginaceae. Family in order Primulales (Bessey), (Hutch.)

Ebenaceae. See 37 4.

Ebenaceae. Ordo in classis Petalanthae (Endl.)

Plantaginaceae. See 41 4.

Plantaginaceae. Family in order Primulales (Bessey).

36 1:1

PLUMBAGINALES

The Plumbaginales include only the Plumbaginaceae. The family is inconspicuous in our area, and the group is not given regular coverage in the present survey. The Plumbaginaceae have pentamerous floral whorls and are classified as gamopetalous and pentacyclic, but they are neither strictly gamopetalous nor strictly pentacyclic. Our forms have petals coherent only at the base. The stamens are epipetalous, and their position opposite the lobes of the corolla is taken as evidence that an outer whorl of stamens has been lost.

The genus Plumbago is not represented in our area. The etymological signification of the name appears to be: "plant with corolla the yellow color of lead oxide." Plumbago as a mineral signified in Latin not "pure lead" (plumbum) but "lead ore," including: "lead oxide" (PbO, yellow), "red lead oxide" (Pb_3O_4, red), and "lead sulfide" (PbS, bluish gray). The ancient use of plumbago as a plant name apparently referred to a plant with yellow corolla. Early modern botanical usage apparently referred it to a plant of the genus Plumbago with bluish-gray corolla. The earliest use of the word plumbago in English refers to red lead oxide. The fact that P. indica has flowers with a red corolla seems, however, only a coincidence.

36 2:1

PLUMBAGINACEAE

dHypoGi̸5L1A5(epipet.opp.)Ci̸5Ki̸5/utr.fl.scarious

Dicotyledonous plants with hypogynous flowers. Gynoecium of 5 united carpels with a single locule. Androecium of 5 stamens (epipetalous and opposite the petals). Corolla of 5 united petals. Calyx of 5 united sepals. Fruit a utricle. Flowers scarious.

PLUMBAGINACEAE

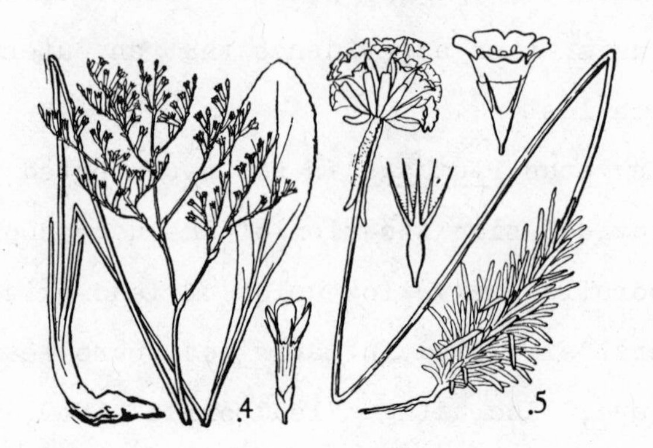

Limonium carolinianum - sea lavender

Armeria labradorica - thrift

Limonium sp.
dHypoG/5L1A5(epipet.opp.)C/5K/5/

36 3:1

Bentham and Hooker, Eichler, Bessey, and Hutchinson do not recognize the Plumbaginales. Engler and Prantl also do not recognize the order, but Engler and Gilg do. In placing the Plumbaginales after the Primulales Engler and Gilg do not disturb the Engler and Prantl sequence of families. Fernald follows the Engler and Gilg system in this detail.

Bentham and Hooker, Eichler, Engler and Prantl, Bessey, and Hutchinson place the Plumbaginaceae in the Primulales. Hutchinson, however, cites with approval the opinion of Bentham and Hooker that the family lacks close affinity with any other. Endlicher places the Plumbaginales at the beginning of the Gamopetalidae, the second of his three dicot subclasses. Here the order stands remote from the Primulales. Within Endlicher's Plumbaginales the Plumbaginaceae follow the Plantaginaceae. Melchior follows the Engler and Gilg system in recognizing the Plumbaginales as following the Primulales, but he disavows the phylogenetic implications in this placement and suggests a derivation of the Plumbaginales from the Caryophyllales. In this suggestion he follows Engler's later views. Cronquist places the Plumbaginales in the Caryophyllidae, the third of his six dicot subclasses. This is remote from his Primulales. He diagrams the Plumbaginaceae as an offshoot on a phylogenetic line from the Phytolaccaceae to the Caryophyllaceae.

There is then a preponderance of opinion to the effect that the Plumbaginales are not closely related to the Primulales. Ultimate derivation of the Plumbaginales from the Caryophyllales is, on the other hand, widely accepted.

GLOSSARY FOR PLUMBAGINALES

Plumbaginales. Order in subclass Gamopetalidae.

Plumbagines. Classis in cohors Gamopetalae (Endl.)

Plumbaginales. Reihe in Unterklasse Sympetalae (Melch.)

Plumbaginales. Order in subclass Caryophyllidae (Cronq.)

Plumbaginaceae. Family in order Plumbaginales.

Plumbagineae. Ordo in classis Plumbaginales (Endl.)

Plumbaginaceae. Ordo in cohors Primulales (B.&H.)

Plumbaginaceae. Familie in Unterreihe Primulinae (Eichl.)

Plumbaginaceae. Familia in series Primulales (Engl.)

Plumbaginaceae. Family in order Primulales (Bessey), (Hutch.)

Plumbaginaceae. Familie in Reihe Plumbaginales (Melch.)

Plumbaginaceae. Family in order Plumbaginales (Cronq.)

EBENALES

Among the families of the Ebenales only the Ebenaceae are here considered. The order is not extensively represented in our area and is not given regular coverage in the present survey. In the Ebenaceae the number of stamens is a multiple of the number of corolla lobes, and the stamens are epipetalous on the gamopetalous corolla.

Latin ebenus is derived from a Greek word for "ebony." This in turn is derived from an ancient Egyptian word. The name Ebenus in botany is pre-Linnaean; but Diospyros is the Linnaean name for the ebony genus, and it signifies etymologically: "seed of Zeus." In ancient Greek the name was applied to the plant now known as Lithospermum officinale. Ancient Greek pyros signifies: "wheat grain."

37 2:1

EBENACEAE

dHypoG\3-nA6-n(epipet.)C_3-7K_3-7/polyg.,dioec.,ber.

Dicotyledonous plants with hypogynous flowers. Gynoecium of 3 to many united carpels. Androecium of 6 to many stamens (epipetalous). Corolla of 3 to 7 petals united at the base. Calyx of 3 to 7 sepals united at the base. Plants polygamous, or dioecious. Fruit a berry.

Diospyros virginiana

dHypoG\4L8A16(m),S8(f),(epipet.)C\4K\4/

Gynoecium of 4 united carpels with a total of 8 locules. Androecium of 16 stamens in staminate flowers, or of 8 staminodia in pistillate flowers, (epipetalous).

EBENACEAE

Diospyros virginiana

persimmon

37 3:1

Endlicher does not recognize the Ebenales but classifies the Ebenaceae in the Primulales. Bentham and Hooker place the Ebenales in the Gamopetalidae, the second of their three dicot subclasses. Here the order is in the superorder Heteromerae: Ericales, Primulales, Ebenales. The name of the superorder alludes to the variability of carpel number within the group. Eichler places the Ebenales in the Gamopetalidae, the first of his two dicot subclasses. Here the order is in the superorder Diplostemones: Primulales, Ebenales. Engler places the Ebenales in the Gamopetalidae, the second of his two dicot subclasses. Here the order is the only member of an order group which has flowers usually with numerous stamens. Bessey places the Ebenales in the Dicotostrobiloididae, the first of his two dicot subclasses. Here the order is in the superorder Sympetalae-Polycarpellatae: Ebenales, Ericales, Primulales. He diagrams as a phylogenetic sequence: Magnoliales, Caryophyllales, Ebenales. Hutchinson places the Ebenales in the Lignosidae, the first of his two dicot subclasses. He diagrams the Ebenales as ultimately derived from the Rhamnales. Melchior classifies the Ebenales in the Gamopetalidae, the second of his two dicot subclasses. Here they stand in a group of orders with morphological similarities: Ericales, Primulales, and Ebenales. Melchior does not, however, recognize phylogenetic relationships between any of these orders. Cronquist places the Ebenales in the Theidae, the fourth of his six dicot subclasses. He diagrams the Ebenales as an offshoot on a line from the Theales to the Primulales.

Although the Ebenales are commonly classified in the same order group as the Ericales and Primulales, they are not regarded as derived from or giving rise to these taxa.

GLOSSARY FOR EBENALES

Ebenales. Order in subclass Gamopetalidae.

Ebenales. Cohors in series Heteromerae (B.&H.) In subclassis Gamopetalae.

Diospyrinae. Unterreihe in Reihe Diplostemones (Eichl.) In Klasse Sympetalae.

Ebenales. Series in subclassis Metachlamydeae (Engl.)

Ebenales. Order in superorder Sympetalae-Polycarpellatae (Bessey). In subclass Strobiloideae.

Ebenales. Order in division Lignosae (Hutch.)

Ebenales. Reihe in Unterklasse Sympetalae (Melch.)

Ebenales. Order in subclass Dilleniidae (Cronq.)

Ebenaceae. Family in order Ebenales.

Ebenaceae. Ordo in classis Petalanthae (Endl.)

Ebenaceae. Ordo in cohors Ebenales (B.&H.)

Ebenaceae. Familie in Unterreihe Diospyrinae (Eichl.)

Ebenaceae. Familia in series Ebenales (Engl.)

Ebenaceae. Family in order Ebenales (Bessey), (Hutch.), (Cronq.)

Ebenaceae. Familie in Reihe Ebenales (Melch.)

TETRACYCLIC GAMOPETALOUS ORDER GROUPS

The orders of Endlicher's subclass Gamopetalidae are arranged in two principal aggregations. The first of these is tetracyclic and includes two order groups. One is epigynous and runs: Asterales, Campanulales, Rubiales. The other is hypogynous and runs: Gentianales, Labiatales, Polemoniales, Scrophulariales. The Plumbaginales, though hypogynous, precede the Asterales. The orders of the Bentham and Hooker subclass Gamopetalidae are distributed into the superorders: Inferae, Heteromerae, and Bicarpellatae. The superorders Inferae and Bicarpellatae are tetracyclic. The Inferae contain the orders: Rubiales, Asterales, and Campanulales. The Bicarpellatae contain the orders: Gentianales, Polemoniales, Scrophulariales, and Labiatales. The orders of Eichler's subclass Gamopetalidae are distributed into the superorders: Haplostemones, Diplostemones, and Obdiplostemones. The superorder Haplostemones is tetracyclic and includes two order groups. One is hypogynous and runs: Polemoniales, Labiatales, Oleales, Gentianales. The other is epigynous and runs: Rubiales, Campanulales. The orders of Engler's subclass are arranged in two aggregations. The second of these is tetracyclic and includes two order groups. One is hypogynous and runs: Gentianales, Polemoniales, Plumbaginales. The other is epigynous and runs: Rubiales, Campanulales. The orders of Bessey's subclass Dicotostrobiloididae are distributed into the superorders: Apopetalae-Polycarpellatae, Sympetalae-Polycarpellatae, and Sympetalae-Dicarpellatae. The superorder Sympetalae-Dicarpellatae is tetracyclic and contains the orders: Gentianales, Polemoniales, Scrophulariales, Labiatales. The orders of Bessey's subclass Dicotocotyloididae are distributed into the superorders: Apopetalae and Sympetalae. The superorder Sympetalae is tetracyclic and contains the orders: Rubiales, Campanulales, Asterales. Hutchinson's dicot subclass Lignosidae contains a tetracyclic order group: Oleales, Apocynales, Rubiales, Bignoniales, and Verbenales. Hutchinson's dicot subclass Herbacidae includes taxa corresponding to portions of the Englerian Polemoniales. These are placed in two order groups. The first contains the Solanales and Scrophulariales. The second contains not only the Polemoniales, Boraginales, and Labiatales but also the Geraniales, from which order Hutchinson derives this last group. Some of the taxa which are commonly given tetracyclic gamopetalous association are treated otherwise by Hutchinson. The Plantaginales are treated as a climax order with derivation from the Primulales, and the Cucurbitales are treated as a climax order with derivation from the Passiflorales. The Caprifoliaceae are treated as a climax family of the climax order Araliales. The Campanulales and Asterales are, however, treated as a gamopetalous, tetracyclic, and epigynous order group. The orders of Melchior's subclass Gamopetalidae in general follow Englerian sequence and are arranged in two aggregations. The second of these is tetracyclic and includes two order groups. One is hypogynous and runs: Oleales, Gentianales, Polemoniales, Plantaginales. The other is epigynous and runs: Caprifoliales, Campanulales. In this system the Rubiaceae are assigned to the Gentianales, and the Cucurbitales are placed after the Violales. Cronquist's dicot subclass Asteridae is tetracyclic and gamopetalous. It includes two order groups. One is hypogynous and runs: Gentianales, Polemoniales, Labiatales, Plantaginales, Scrophulariales. The other is epigynous and runs: Campanulales, Rubiales, Caprifoliales, Asterales. In this system the Cucurbitaceae are assigned to the Violales.

38 Ob:1

GLOSSARY FOR TETRACYCLIC GAMOPETALOUS SUPERORDERS, ETC.

Inferae (cohortes: Rubiales, Asterales, Campanales). Series in subclassis Gamopetalae (B.&H.)

Bicarpellatae (cohortes: Gentianales, Polemoniales,

Personales, Lamiales). Series in subclassis Gamopetalae (B.&H.)

Haplostemones (Unterreihen: Tubiflorae, Labiatiflorae, Ligustrinae, Contortae, Aggregatae, Campanulinae). Reihe in Klasse Sympetalae (Eichl.)

Sympetalae-Dicarpellatae (orders: Gentianales, Polemoniales, Scrophulariales, Lamiales). Superorder in subclass Strobiloideae (Bessey).

Sympetalae (orders: Rubiales, Campanulales, Asterales). Superorder in subclass Cotyloideae (Bessey).

38 1:1

OLEALES

Among the families of the Oleales only the Oleaceae are represented in our area. The Oleaceae have flowers gamopetalous and tetracyclic. Their gynoecium is composed of two united carpels, and their androecium has stamens usually isomeric with the carpels. Included in the Oleaceae are the genera _Fraxinus_, _Forsythia_, _Ligustrum_, and _Syringa_.

The genus Olea does not occur in our area. In ancient Latin the name _olea_ designated the olive fruit and hence also the olive tree. _Oleum_, "olive oil" and hence simply "oil," is the source of the English term _oil_.

OLEACEAE

Fraxinus pennsylvanica	Ligustrum vulgare	Syringa vulgaris
-	-	-
ash	privet	lilac

OLEACEAE

dHypoG/2A2-4(epipet.alt.)C/4,O,K,P,/4/perf.,polyg.,dioec., caps.,sam.,drup.,lvs.opp.

Dicotyledonous plants with hypogynous flowers. Gynoecium of 2 united carpels. Androecium of 2 to 4 stamens (epipetalous and alternate with the petals, if any). Corolla of 4 united petals, or none. Calyx, or perianth, of 4 united sepals, or tepals. Flowers perfect, or plants polygamous or dioecious. Fruit a capsule, or samara, or drupaceous. Leaves opposite.

Forsythia sp.

dHypoGi̸2A2(epipet.alt.)Ci̸4Ki̸4/

38 3:1

Endlicher, Bentham and Hooker, Engler, and Bessey do not recognize the Oleales but place the Oleaceae in the Gentianales. Cronquist also does not recognize the Oleales, but he alone places the Oleaceae in the Scrophulariales. Eichler places the Oleales in the Gamopetalidae, the first of his two dicot subclasses. Here they stand in a tetracyclic superorder Haplostemones and within this in the hypogynous order group: Polemoniales, Labiatales, Oleales, Gentianales. Hutchinson places the Oleales and orders derivative from them at the end of the Lignosidae, the first of his two dicot subclasses. He diagrams as a phylogenetic sequence: Magnoliales, Oleales. Here the Oleaceae are interpreted as a highly modified climax group of an extensive order. Melchior places the Oleales in the Gamopetalidae, the second of his two dicot subclasses. Here the order is associated in the tetracyclic and hypogynous order group: Oleales, Gentianales, Polemoniales, Plantaginales. This sequence had previously been adopted by Fernald in the eighth edition of Gray's Manual.

Current systems reject the incorporation of the Oleaceae into the Gentianales, but they differ among themselves in their classifications of the family.

GLOSSARY FOR OLEALES

Oleales. Order in subclass Gamopetalidae.

  Ligustrinae. Unterreihe in Reihe Haplostemones (Eichl.) In Klasse Sympetalae.

  Loganiales. Order in division Lignosae (Hutch.)

  Oleales. Reihe in Unterklasse Sympetalae (Melch.)

Oleaceae. Family in order Oleales.

  Oleaceae. Ordo in classis Contortae (Endl.)

  Oleaceae. Ordo in cohors Gentianales (B.&H.)

  Oleaceae. Familie in Unterreihe Ligustrinae (Eichl.)

  Oleaceae. Familia in series Contortae (Engl.)

  Oleaceae. Family in order Gentianales (Bessey).

  Oleaceae. Family in order Loganiales (Hutch.)

  Oleaceae. Familie in Reihe Oleales (Melch.)

  Oleaceae. Family in order Scrophulariales (Cronq.)

GENTIANALES

The Gentianales contain the families Gentianaceae, Apocynaceae, and Asclepiadaceae. The order has also the descriptive name Contortae in allusion to corolla segments with usually twisted aestivation. The order is gamopetalous, tetracyclic, and bicarpellate. Leaves are usually opposite.

The name *Gentiana* is derived through ancient Latin from a Greek name for the gentian. The etymological signification of the name is unknown. *Apocynum* is derived through ancient Latin from a Greek word signifying: "plant which puts dogs out of the way"; i.e., kills them. Various plants were anciently used to poison dog food, and these same plants were given other names when used as poisons for other animals. *Asclepias* is derived through ancient Latin from a Greek name for swallowwort. This plant, formerly classified in the modern genus *Asclepias*, is now classified in another genus of the Aslepiadaceae. *Asclepias* signifies: "plant sacred to Asclepius," a mythical physician.

39 2:1

GENTIANACEAE

dHypoGi̸2L1A4-5(epipet.alt.)Ci̸4-5Ki̸4-5/caps.aestiv.contort.

lvs.opp.

GENTIANACEAE

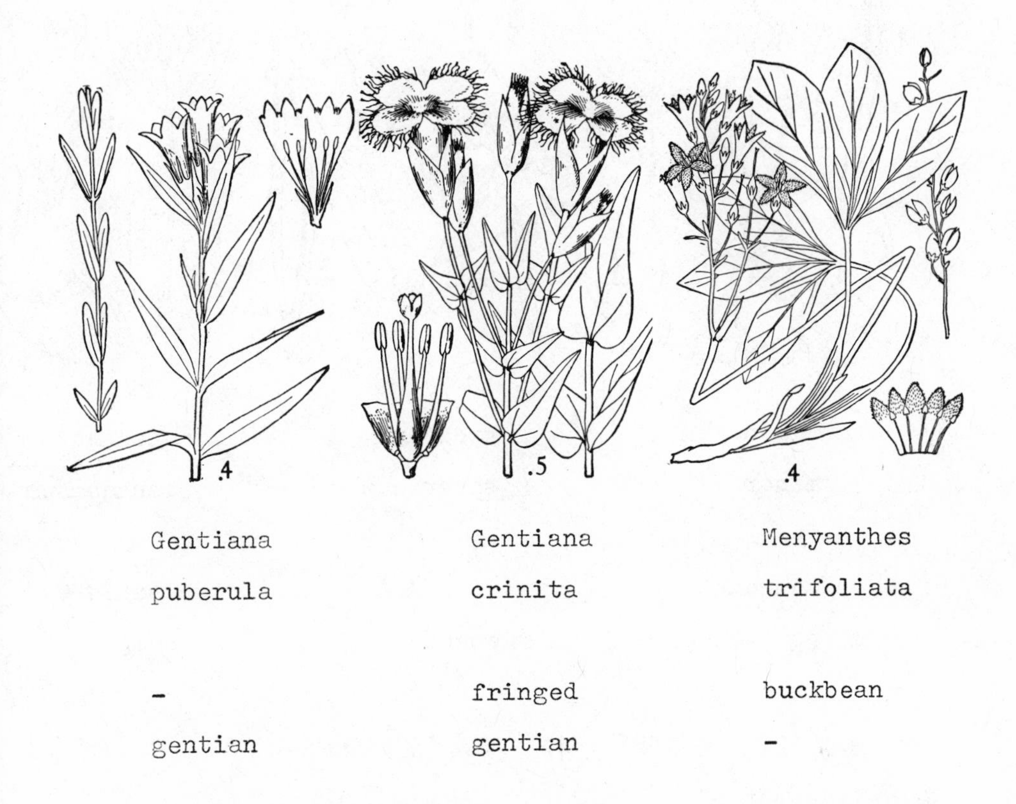

Gentiana puberula	Gentiana crinita	Menyanthes trifoliata
- gentian	fringed gentian	buckbean -

Dicotyledonous plants with hypogynous flowers. Gynoecium of 2 united carpels and a single locule. Androecium of 4 to 5 stamens (epipetalous and alternate with the petals). Corolla of 4 to 5 united petals. Calyx of 4 to 5 united sepals. Fruit a capsule. Aestivation (of corolla) contorted. Leaves opposite.

Gentiana sp.

dHypoG✗2L1A5(epipet.alt.)C✗5K✗5/

APOCYNACEAE

Vinca minor	Apocynum sibericum	Apocynum cannabinum
common periwinkle	- dogbane	Indian hemp

APOCYNACEAE

dHypoGi̸2(L2,1),i2(w.styles i̸ into clavuncle),A5(epipet. alt.)Ci̸5K_5/fol./juice milky

Dicotyledonous plants with hypogynous flowers. Gynoecium of 2 united carpels (and 2 locules, or a single locule), or of 2 distinct carpels (with styles united into a clavuncle). Androecium of 5 stamens (epipetalous and alternate with the petals). Corolla of 5 united petals. Calyx of 5 sepals united at the base. Fruit a follicle. Juice milky.

Apocynum sp.

dHypoG2(w.styles / into clavuncle)A5(epipet.alt.)C/5K/5/

ASCLEPIADACEAE

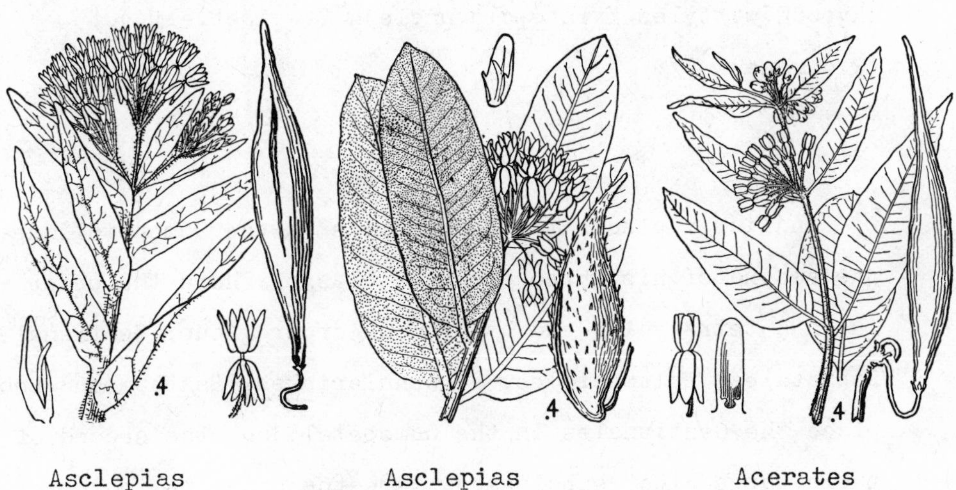

Asclepias tuberosa	Asclepias syriaca	Acerates viridiflora
butterfly-weed	common milkweed	green milkweed

ASCLEPIADACEAE

dHypoG2(w.styles / into clavuncle)A¯5(epipet.alt.)C/5 (oft.w.corona)K_5/fol./juice milky

Dicotyledonous plants with hypogynous flowers. Gynoecium of 2 carpels (with styles united into a clavuncle). Androecium of 5 stamens united by the anthers (epipetalous and alternate with the petals. Corolla of 5 united petals (often with a corona). Calyx of 5 sepals united at the base. Fruit a follicle. Juice milky.

Asclepias sp.

dHypoG2(w.styles / into clavuncle)A¯5(epipet.alt.)C/5 (w.corona)K_5/

39 3:1

Endlicher places the Gentianales in the Gamopetalidae, the second of his three dicot subclasses. Here the order is in a tetracyclic and hypogynous order group: Gentianales, Labiatales, Polemoniales, Scrophulariales. Bentham and Hooke place the Gentianales in the Gamopetalidae, the second of their three dicot subclasses. Here the order is in the superorder Bicarpellatae: Gentianales, Polemoniales, Scrophu- lariales, Labiatales. The Plantaginaceae are also referred to this group. Here the family is treated as anomalous and is assigned to no order. This is equivalent to the creation of an order Plantaginales for this single family. Eichler places the Gentianales in the Gamopetalidae, the first of his two dicot subclasses. Here the order is in the superorde Haplostemones, and within this in a hypogynous order group: Polemoniales, Labiatales, Oleales, Gentianales. Engler place the Gentianales in the Gamopetalidae, the second of his two dicot subclasses. Here the order is in the tetracyclic and hypogynous order group: Gentianales, Polemoniales, Plantaginales. Bessey places the Gentianales in the Dicotostrobiloididae, the first of his two dicot subclasses. Here the order is in the superorder Sympetalae-Dicarpellatae: Gentianales, Polemoniales, Scrophulariales, Labiatales. He diagrams as a phylogenetic sequence: Magnoliales, Caryophyllales, Primulales, Gentianales. Hutchinson places the Gentianales in the Herbacidae, the second of his two dicot subclasses. He diagrams as a phylogenetic sequence: Ranunculales, Caryophyllales, Gentianales. Melchior maintains the basic Englerian position for the Gentianales. Within the Gamopetalidae they are in the tetracyclic and hypogynous order group: Oleales, Gentianales, Polemoniales, Plantaginales. Cronquist places the Gentianales in the subclass Asteridae, the sixth of his six dicot subclasses. Here the order is in a hypogynous order group: Gentianales, Labiatales, Plantaginales, Polemoniales, Scrophulariales.

Endlicher's and Bentham and Hooker's families within the Gentianales have the sequence: Oleaceae, Apocynaceae, Asclepiadaceae, Gentianaceae. Eichler and Cronquist, however, give the Gentianaceae first position with the sequence: Gentianaceae, Apocynaceae, Asclepiadaceae. Engler and Bessey add the Oleaceae to produce the sequence: Oleaceae, Gentianaceae, Apocynaceae, Asclepiadaceae. Of the families here considered, Hutchinson includes only the Gentianaceae in the order Gentianales. He places the Apocynaceae and

Asclepiadaceae in the subclass Lignosidae within the order Apocynales. Melchior alone includes the Rubiaceae in the order with the sequence: Gentianaceae, Apocynaceae, Asclepiadaceae, Rubiaceae.

There is consensus (Hutchinson dissenting) associating the Gentianales and Polemoniales in an order group. Within the Gentianales there is in later systems consensus (Hutchinson dissenting) favoring the sequence: Gentianaceae, Apocynaceae, Asclepiadaceae.

39 4:1

GLOSSARY FOR GENTIANALES

Gentianales. Order in subclass Gamopetalidae.

Contortae. Classis in cohors Gamopetalae (Endl.)

Gentianales. Cohors in series Bicarpellatae (B.&H.) In subclass Gamopetalae.

Contortae. Unterreihe in Reihe Haplostemones (Eichl.) In Klasse Sympetalae.

Contortae. Series in subclassis Metachlamydeae (Engl.)

Gentianales. Order in superorder Sympetalae-Dicarpellatae (Bessey). In subclass Strobiloideae.

Gentianales. Order in division Herbaceae (Hutch.)

Gentianales. Reihe in Unterklasse Sympetalae (Melch.)

Gentianales. Order in subclass Asteridae (Cronq.)

Apocynales. Order in subclass Gamopetalidae.

Apocynales. Order in division Lignosae (Hutch.)

Gentianaceae. Family in order Gentianales.

Gentianeae. Ordo in classis Contortae (Endl.)

Gentianeae. Ordo in cohors Gentianales (B.&H.)

Gentianaceae. Familie in Unterreihe Contortae (Eichl.)

Gentianaceae. Familia in series Contortae (Engl.)

Gentianaceae. Family in order Gentianales (Bessey), (Hutch.), (Cronq.)

Gentianaceae. Familie in Reihe Gentianales (Melch.)

Apocynaceae. Family in order Gentianales.

Apocynaceae. Ordo in classis Contortae (Endl.)

Apocynaceae. Ordo in cohors Gentianales (B.&H.)

Apocynaceae. Familie in Unterreihe Contortae (Eichl.)

Apocynaceae. Familia in series Contortae (Engl.)

Apocynaceae. Family in order Gentianales (Bessey), (Cronq.)

Apocynaceae. Family in order Apocynales (Hutch.)

Apocynaceae. Familie in Reihe Gentianales (Melch.)

Asclepiadaceae. Family in order Gentianales.

Asclepiadeae. Ordo in classis Contortae (Endl.)

Asclepiadeae. Ordo in cohors Gentianales (B.&H.)

Asclepiadaceae. Familie in Unterreihe Contortae (Eichl.)

Asclepiadaceae. Familia in series Contortae (Engl.)

Asclepiadaceae. Family in order Gentianales (Bessey), (Cronq.)

Asclepiadaceae. Family in order Apocynales (Hutch.)

Asclepiadaceae. Familie in Reihe Gentianales (Melch.)

Oleaceae. See 38 4.

Oleaceae. Ordo in classis Contortae (Endl.)

Oleaceae. Ordo in cohors Gentianales (B.&H.)

Oleaceae. Familia in series Contortae (Engl.)

Oleaceae. Family in order Gentianales (Bessey).

Rubiaceae. See 42 4.

Rubiaceae. Familie in Reihe Gentianales (Melch.)

40 1:1

POLEMONIALES

The Polemoniales are gamopetalous, tetracyclic, and hypogynous. Their flowers have united carpels and have the corollae mostly not contorted in aestivation. The order includes the Convolvulaceae, Polemoniaceae, Hydrophyllaceae, Boraginaceae, Verbenaceae, Labiatae, Solanaceae, Scrophulariaceae, and Bignoniaceae. The Convolvulaceae have flowers with two carpels. Their corollae are contorted in aestivation, but tubular or salverform at maturity. The plants are herbaceous and vining. The Polemoniaceae have flowers with three carpels. Their corollae are salverform or campanulate. In our area the plants of this family are herbaceous. The

Hydrophyllaceae have flowers with two carpels but with a single locule, or two locules formed by fusion of placentae. The plants are herbaceous. The Boraginaceae have flowers with two carpels, but their pistil is four-locular at maturity. The inflorescences of the family are commonly circinate. The plants are herbaceous. The Verbenaceae have flowers with two carpels and four locules; the style is terminal; the corolla is zygomorphic. The plants are herbaceous or shrubby. The Labiatae have flowers with two carpels and four locules; the style is usually gynobasic and two-cleft at the tip; the corolla is zygomorphic. The plants are herbaceous. The Solanaceae have flowers with two carpels placed obliquely with respect to the floral axis. The plants are herbaceous or shrubby. This family includes several species of importance either for food or medicine. The Scrophulariaceae have flowers with two carpels; the corolla is usually zygomorphic. The plants in our area are herbs. The Bignoniaceae have flowers with two carpels, and in our species two locules; the corolla is zygomorphic, with the tube longer than the lobes. The plants are trees and vines.

Convolvulus is the ancient Latin name for the bindweed, and signifies: "plant with tendrils which roll up." Polemonium is derived from an ancient Greek name applied to various plants, and its etymological signification is obscure. Hydrophyllum is a pre-Linnaean name derived from

Greek roots to signify: "plant with leaves holding water," although only the petiole portions are watery. *Borago* is the medieval Latin name of borage, but its etymological signification is in doubt. *Verbena* is an ancient Latin term for ceremonial foliage rather than a name for a specific class of plants. In post-classical times the name became applied to vervain, probably in connection with medicinal uses. *Labiatae* is a descriptive name signifying: "family having a two-lipped corolla." *Solanum* is an ancient Latin name applied to plants of the nightshade group used medicinally. If the element *sol-* may be compared with such words as *solacium*, "solace" or "comfort," and *solamen*, "solace," then the name *Solanum* may be interpreted to signify: "plant which is employed medicinally to relieve pain." *Scrophularia* is a medieval Latin name for a medicine made from figwort leaves and used in the treatment of scrophula. Hence the name was extended to signify: "plant from which the medicine for scrophula is derived." *Bignonia* is a name created by Tournefort to signify: "genus commemorative of the Abbé Bignon," librarian of Louis XIV.

40 2:1

CONVOLVULACEAE

dHypoGi̸2A5(epipet.alt.)Ci̸5Ki,_,5/caps.

Dicotyledonous plants with hypogynous flowers. Gynoecium of 2 united carpels. Androecium of 5 stamens

CONVOLVULACEAE

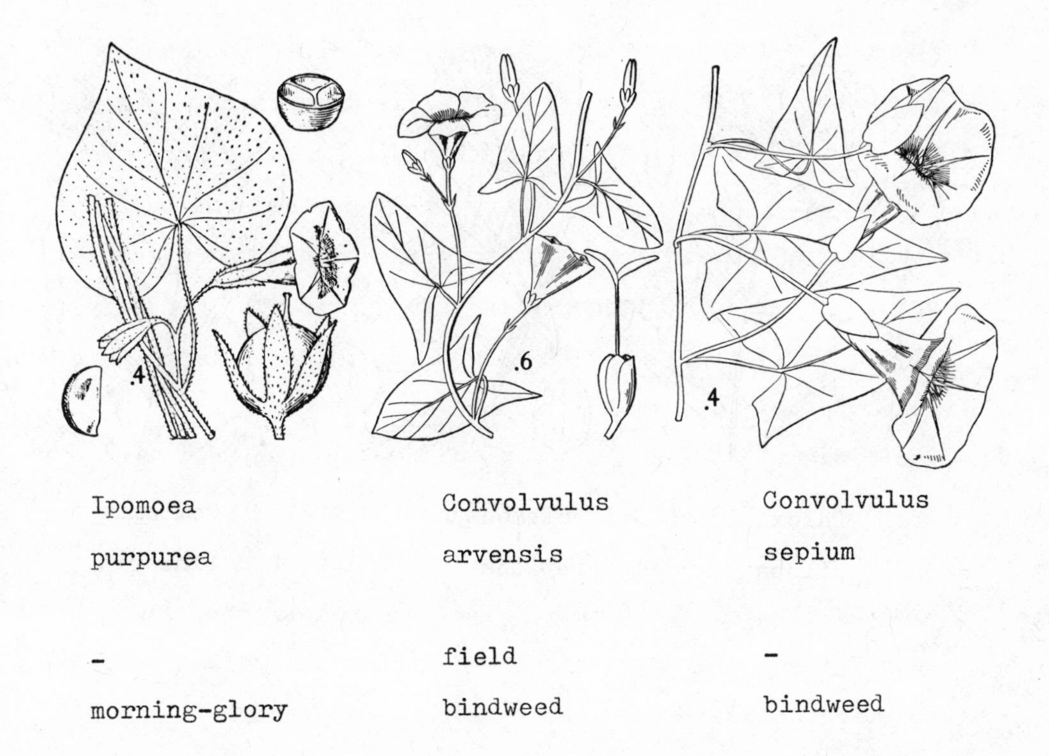

Ipomoea purpurea	Convolvulus arvensis	Convolvulus sepium
- morning-glory	field bindweed	- bindweed

(epipetalous and alternate with the petals). Corolla of 5 united petals. Calyx of 5 sepals, free or united at the base. Fruit a capsule.

Convolvulus sp.

dHypoG/2A5(epipet.alt.)C/5K5/

POLEMONIACEAE

dHypoG/3A5(epipet.alt.)C/5K/5/caps.

POLEMONIACEAE

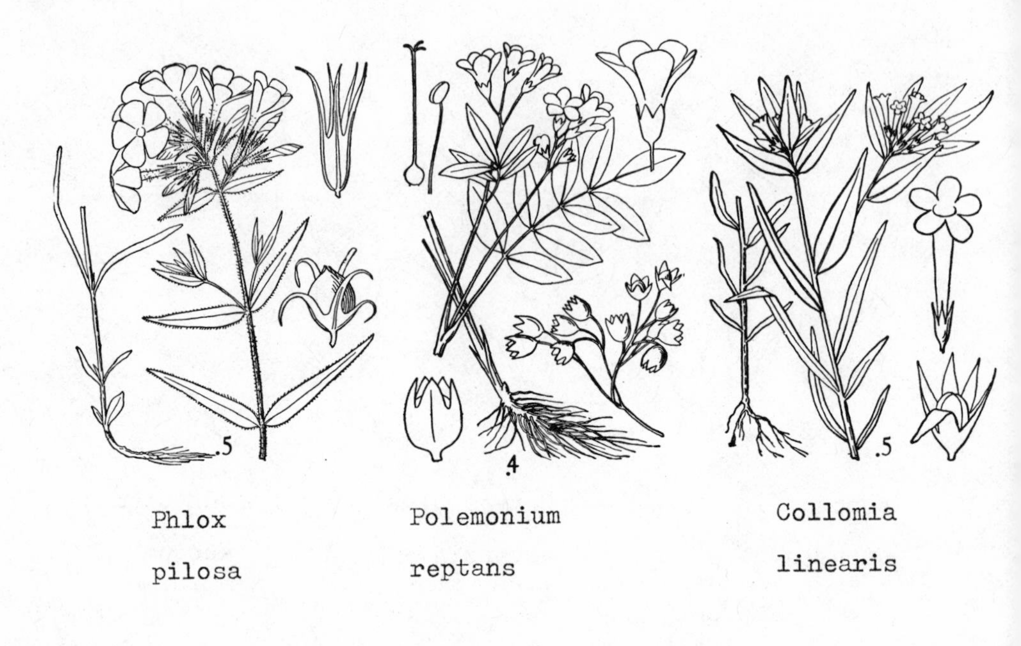

Phlox pilosa | Polemonium reptans | Collomia linearis

Dicotyledonous plants with hypogynous flowers. Gynoecium of 3 united carpels. Androecium of 5 stamens (epipetalous and alternate with the petals). Corolla of 5 united petals. Calyx of 5 united sepals. Fruit a capsule.

Phlox sp.

dHypoG/3A5(epipet.alt.)C/5K/5/

HYDROPHYLLACEAE

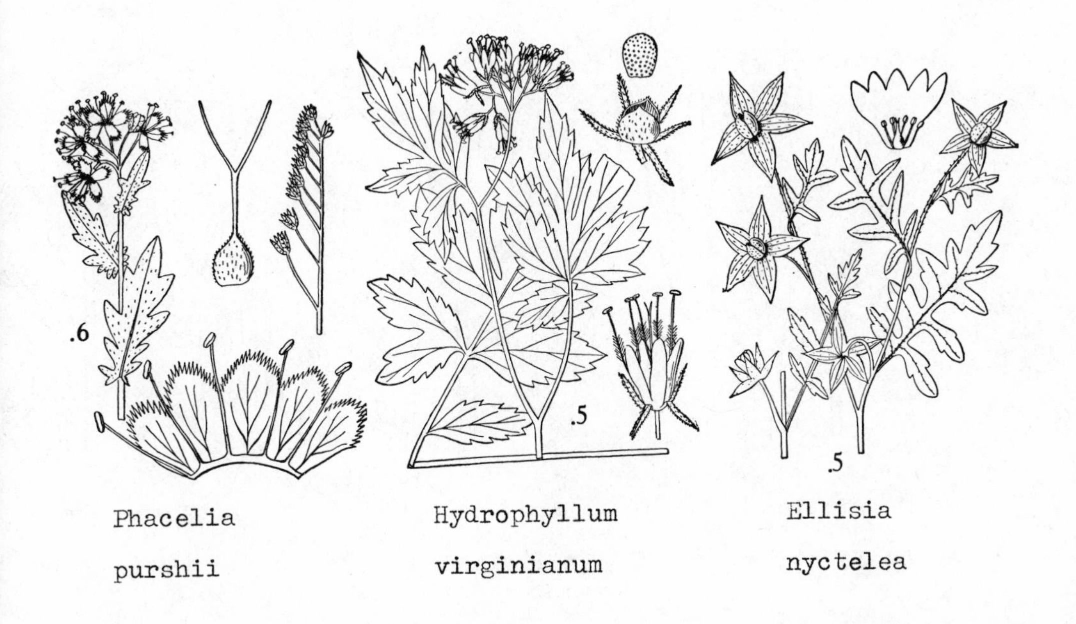

Phacelia purshii — Hydrophyllum virginianum — Ellisia nyctelea

\- \- \-

HYDROPHYLLACEAE

dHypoG̸2L1,2,A5(epipet.alt.)C̸5K_5/caps.

Dicotyledonous plants with hypogynous flowers. Gynoecium of 2 united carpels and a single locule, or 2. Androecium of 5 stamens (epipetalous and alternate with the petals). Corolla of 5 united petals. Calyx of 5 sepals united at the base. Fruit a capsule.

Hydrophyllum sp.

dHypoG̸2L1A5(epipet.alt.)C̸5K_5/

BORAGINACEAE

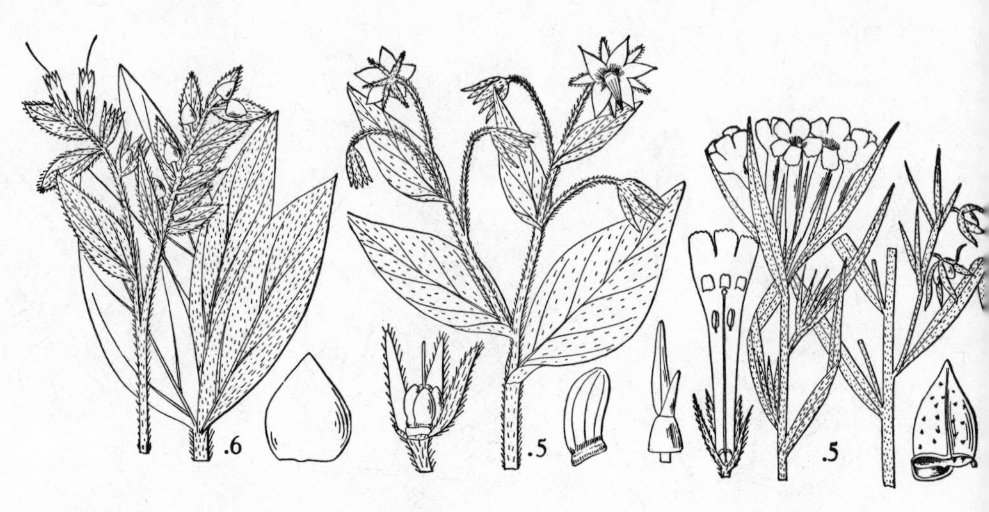

Onosmodium occidentale	Borago officinalis	Lithospermum incisum
-	-	-
	borage	

BORAGINACEAE

dHypoG$\not{2}$(4-lobed)L4(at matur.)A5(epipet.alt.)C$\not{5}$K$^{\pm}\not{5}$/ nutlets 1-4

Dicotyledonous plants with hypogynous flowers. Gynoecium of 2 united carpels (4-lobed) and 4 locules (at maturity). Androecium of 5 stamens (epipetalous and alternate with the petals). Corolla of 5 united petals. Calyx of 5 more or less united sepals. Fruit 1 to 4 nutlets.

Borago officinalis

dHypoG/2(4-lobed)L4(at matur.)A5(epipet.alt.)C/5K_5/

VERBENACEAE

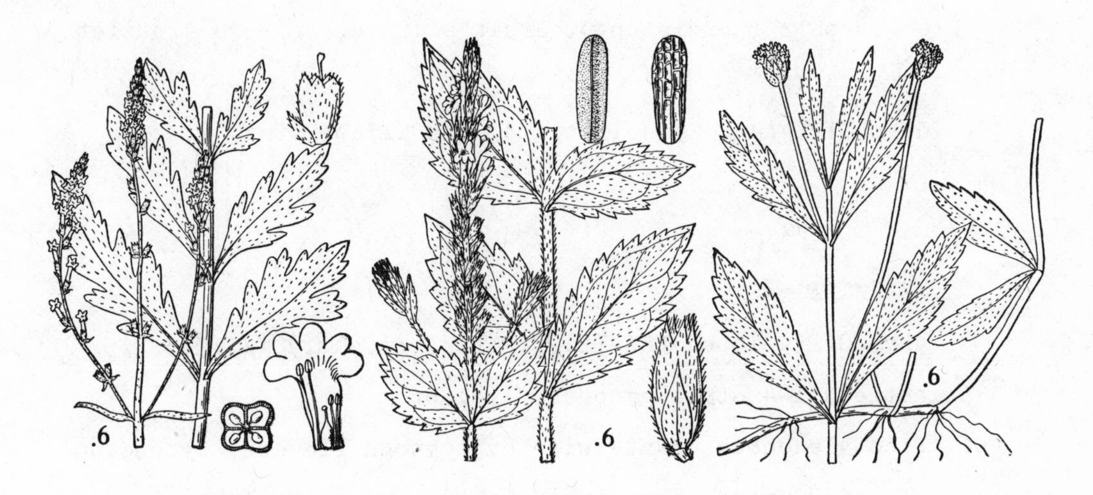

Verbena officinalis	Verbena stricta	Lippia lanceolata
-	-	-
vervain	vervain	fog-fruit

VERBENACEAE

dHypoG/2L2-4A4(2+2)(epipet.alt.)C/ø4-5K/4-5(persist.)/

drup.,nutlets 2-4

Dicotyledonous plants with hypogynous flowers. Gynoecium of 2 united carpels and 2 to 4 locules. Androecium of 4 stamens (two sets of 2 stamens each) (epipetalous and alternate with the petals). Corolla of 4 to 5 petals, united and zygomorphic. Calyx of 4 to 5 united sepals (persistent). Fruit a drupe, or 2 to 4 nutlets.

Verbena sp.

dHypoGi̸2L4A4(2+2)(epipet.alt.)Ci̸4̸5Ki̸5(persist.)/

LABIATAE

dHypoGi̸2(4-lobed)L4A4(2+2),2,(epipet.alt.)Ci̸4̸5(2+3)Ki̸5/
nutlets 1-4 style gynobas./lvs.opp.

Dicotyledonous plants with hypogynous flowers. Gynoecium of 2 united carpels (4-lobed) and 4 locules. Androecium of 4 stamens (two sets of 2 stamens each), or of only 2 stamens, (epipetalous and alternate with the petals). Corolla of 5 petals, united and zygomorphic (the lobes in groups of 2 and of 3 members). Calyx of 5 united sepals. Fruit of 1 to 4 nutlets. Style gynobasic. Leaves opposite.

Lamium sp.

dHypoGi̸2(4-lobed)L4A4(2+2)(epipet.alt.)Ci̸4̸5(2+3)Ki̸5/

LABIATAE

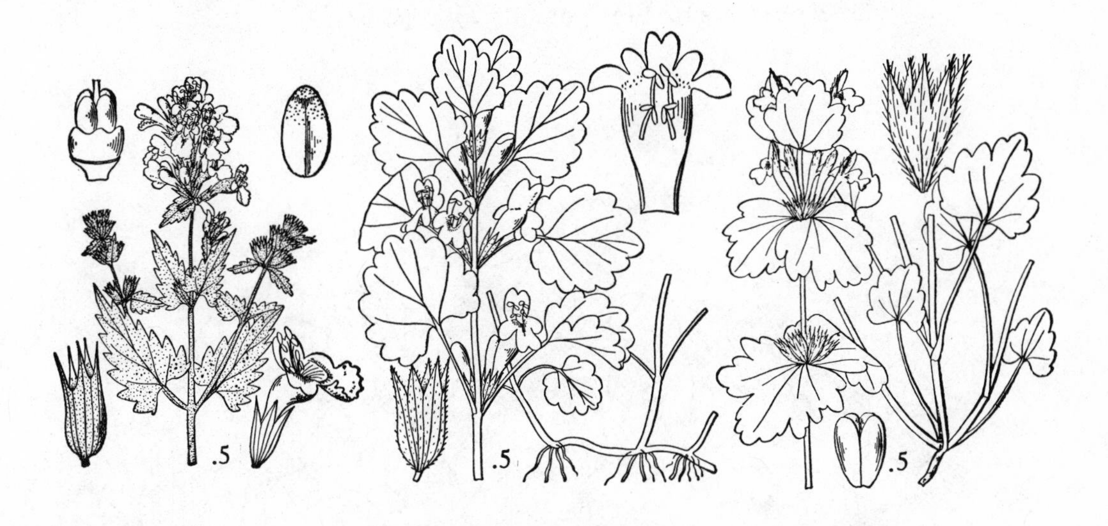

Nepeta cataria	Glecoma hederacea	Lamium amplexicaule
catnip	-	henbit
-	ground ivy	-

SOLANACEAE

dHypoG/2(oblique)A5(epipet.alt.)C/a,/a,5K/5/ber.,caps.

Dicotyledonous plants with hypogynous flowers. Gynoecium of 2 united carpels (oblique). Androecium of 5 stamens (epipetalous and alternate with the petals). Corolla of 5 petals, united and actinomorphic or zygomorphic. Calyx of 5 united sepals. Fruit a berry, or capsule.

Solanum sp.

dHypoG/2(oblique)A5(epipet.alt.)C/5K/5/

SOLANACEAE

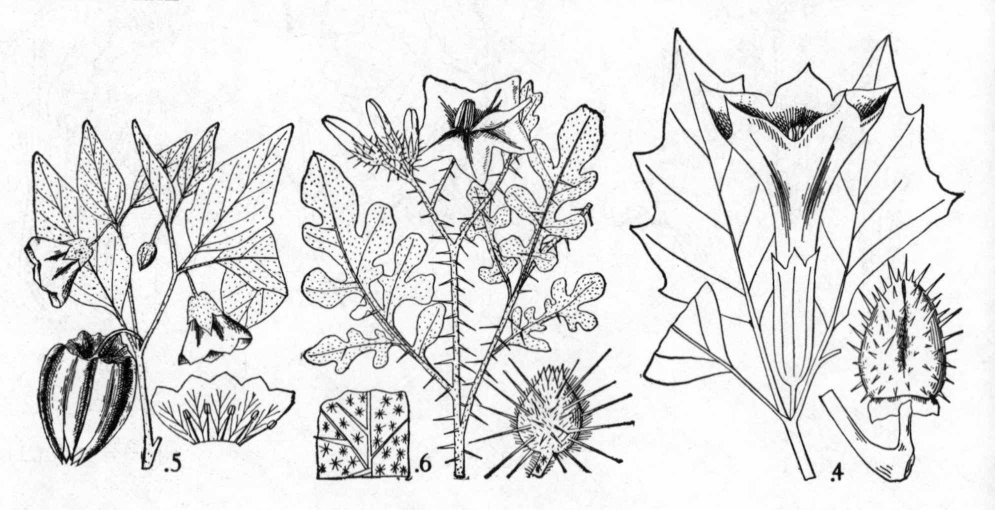

Physalis virginiana	Solanum rostratum	Datura stramonium
-	buffalo-bur	-
ground cherry	-	Jimsonweed

SCROPHULARIACEAE

dHypoG/2A(2-)4(-5)(oft.S1,SO,+2+2)(epipet.alt.)C/ø,a,
5(2+3)K/5/caps.herb.

SCROPHULARIACEAE

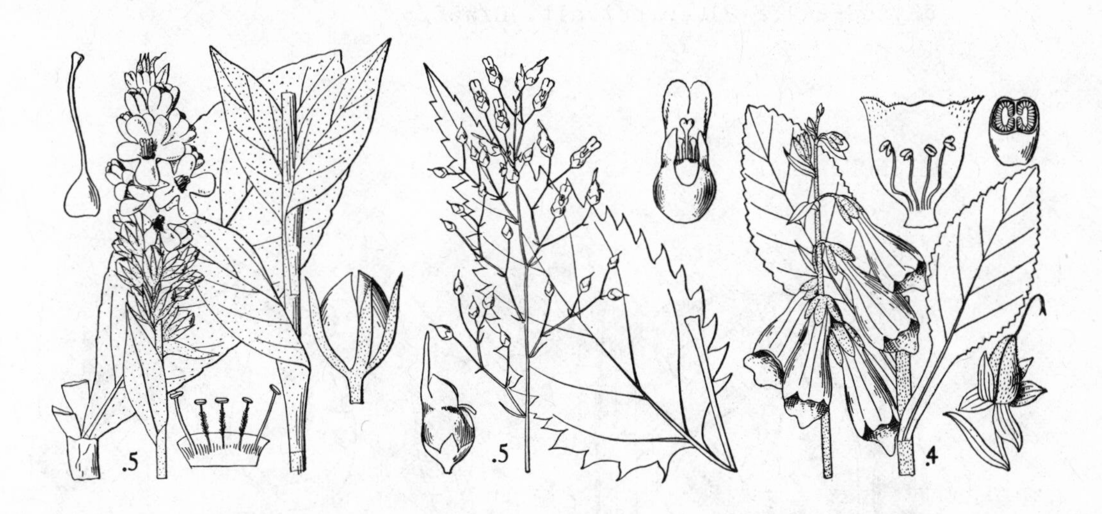

Verbascum thapsus	Scrophularia lanceolata	Digitalis purpurea
-	-	-
mullein	figwort	foxglove

Dicotyledonous plants with hypogynous flowers. Gynoecium of 2 united carpels. Androecium of 4 stamens (or as few as 2 or as many as 5) (often a single staminodium, or none, and two sets of 2 stamens each) (the stamens epipetalous and alternate with the petals). Corolla of 5 petals, united and zygomorphic or actinomorphic, (the lobes in groups of 2 and of 3 members). Calyx of 5 united sepals. Fruit a capsule. Plants herbaceous.

Digitalis sp.

dHypoG/2A4(2+2)(epipet.alt.)C/ø5K/5/

BIGNONIACEAE

Campsis radicans	Bignonia capreolata	Catalpa speciosa
-	cross vine	-
trumpet creeper	-	

BIGNONIACEAE

dHypoG/2A4(2+2),2,(epipet.alt.)C/ø5K/5,2/caps./woody/lvs.opp.

Dicotyledonous plants with hypogynous flowers. Gynoecium of 2 united carpels. Androecium of 4 stamens (two sets of 2 stamens each), or stamens 2, (the stamens epipetalous and alternate with the petals). Corolla of 5 petals, united and zygomorphic. Calyx of 5, or 2, united sepals. Fruit a capsule. Plants woody. Leaves opposite.

Campsis radicans

dHypoGi̸2A4(2+2)(epipet.alt.)Ci̸a̸5Ki̸5/

40 3:1

Endlicher recognizes the orders Labiatales, Polemoniales, Scrophulariales; and he places them in the Gamopetalidae, the second of his three dicot subclasses. Here the sequence is in the tetracyclic and hypogynous order group: Gentianales, Labiatales, Polemoniales, Scrophulariales. Bentham and Hooker recognize the orders Polemoniales, Scrophulariales, Labiatales; and they place them in the Gamopetalidae, the second of their three dicot subclasses. Here the sequence is in the tetracyclic and hypogynous superorder Bicarpellatae: Gentianales, Polemoniales, Scrophulariales, Labiatales. The Plantaginaceae are also referred to this group. Here the family is treated as anomalous and is assigned to no order. This is equivalent to the creation of an order Plantaginales for this single family. Eichler recognizes the orders Polemoniales (incl.

Scrophulariaceae), Labiatales; and he places them in the Gamopetalidae, the first of his two dicot subclasses. Here the sequence is in the tetracyclic superorder Haplostemones and in the hypogynous order group: Polemoniales, Labiatales, Oleales, Gentianales. Engler recognizes the Polemoniales (incl. Scrophulariaceae and Labiatae); and he places them in the Gamopetalidae, the second of his two dicot subclasses. Here the order is in the tetracyclic and hypogynous order group: Gentianales, Polemoniales, Plantaginales. Bessey, following the Bentham and Hooker sequence, recognizes the Polemoniales, Scrophulariales, and Labiatales. He places the sequence in the Dicotostrobiloididae, the first of his two dicot subclasses. Here the sequence is in the tetracyclic and hypogynous superorder Sympetalae-Dicarpellatae: Gentianales, Polemoniales, Scrophulariales, Labiatales. He diagrams the Gentianales and Polemoniales as separately derived from the Primulales, and he also diagrams the Scrophulariales and Labiatales as separately derived from the Polemoniales. At the end of the Herbacidae, the second of his two dicot subclasses Hutchinson recognizes two predominantly tetracyclic and hypogynous order groups. One contains the Solanales and Scrophulariales. The other contains the Geraniales, Polemoniales, Boraginales, and Labiatales. He diagrams as one phylogenetic sequence: Saxifragales, Solanales, Scrophulariales, and he diagrams as another phylogenetic sequence: Saxifragales, Geraniales, Polemoniales, Boraginales, Labiatales. He diagrams the Apocynales, Rubiales, Bignoniales, and Verbenales as separately derived from the Oleales (but not from the Oleaceae). Melchior maintains the Englerian concept of the Polemoniales (incl. Scrophulariaceae and Labiatae), and he follows the Englerian placement of the order. Cronquist recognizes the orders Labiatales, Polemoniales, Scrophulariales; and he places them in the gamopetalous and tetracyclic Asteridae, the sixth of his six dicot subclasses. Here they are in the hypogynous order group: Gentianales, Polemoniales, Labiatales, Plantaginales, Scrophulariales. Following Bessey, he diagrams the Scrophulariales and Labiatales as separately derived from the Polemoniales.

There is then consensus (Engler and Melchior dissenting) that the Labiatales should be recognized at ordinal rank, and usually the Polemoniales, Labiatales, and Scrophulariales are treated as coordinate related taxa. Engler and Melchior recognize suborders which may be compared with these orders of the other systems.

Endlicher's Polemoniales contain the families: Convolvulaceae, Polemoniaceae, Hydrophyllaceae, and Solanaceae. Bentham and Hooker's Polemoniales contain the Polemoniaceae, Hydrophyllaceae, Boraginaceae, Convolvulaceae, and Solanaceae. Eichler rearranges the Bentham and Hooker families to run: Convolvulaceae, Polemoniaceae, Hydrophyllaceae, Boraginaceae, and Solanaceae. Engler's Polemonium suborder runs: Convolvulaceae, Polemoniaceae; and his Borago suborder runs: Hydrophyllaceae, Boraginaceae. The two suborders taken together are comparable to the Eichlerian order (without the Solanaceae). Bessey's Polemoniales contain the same families as the orders of Bentham and Hooker and Eichler, but the sequence is: Polemoniaceae, Convolvulaceae, Hydrophyllaceae, Boraginaceae, and Solanaceae. Hutchinson's Polemoniales contain the families: Polemoniaceae, Hydrophyllaceae. His Solanales contain the Solanaceae and Convolvulaceae. These two orders taken together are comparable to the Endlicher order, but in Hutchinson's system they are not directly related. The Polemonium and Borago suborders of Melchior correspond to those of Engler, but Melchior places the Polemoniaceae ahead of the Convolvulaceae. Cronquist's sequence for the Polemoniales runs: Solanaceae, Convolvulaceae, Polemoniaceae, Hydrophyllaceae. These families correspond with those of Endlicher's order.

Endlicher's Labiatales contain the Labiatae, Verbenaceae, and Boraginaceae. Bentham and Hooker's Labiatales contain the Verbenaceae and Labiatae. Eichler's Labiatales contain the Scrophulariaceae, Bignoniaceae, Plantaginaceae, Verbenaceae, and Labiatae. Engler's Lamium suborder contains the Verbenaceae and Labiatae. This suborder is comparable to the order of Bentham and Hooker. Bessey's Labiatales contain the Verbenaceae and Labiatae. Hutchinson's Labiatales contain the Labiatae. As previously noted, his Verbenaceae are in a different subclass from that of his Labiatae. The *Lamium* suborder of Melchior corresponds to that of Engler. Cronquist's Labiatales contains the families: Boraginaceae, Verbenaceae, and Labiatae. These families correspond with those of Endlicher's order.

The Scrophulariales of Endlicher, Bentham and Hooker, and Bessey contain the Scrophulariaceae and Bignoniaceae. These families as just noted are included in Eichler's Labiatales. The *Scrophularia* suborder of Engler contains the Solanaceae, Scrophulariaceae, and Bignoniaceae. Hutchinson's Scrophulariales contain only the Scrophulariaceae. His Bignoniaceae are in the Bignoniales which, as previously noted, are in the subclass Lignosidae. The *Scrophularia* suborder of Melchior corresponds to that of Engler. Cronquist's Scrophulariales contain the Oleaceae, Scrophulariaceae, and Bignoniaceae.

There is then consensus that the Polemoniaceae and Hydrophyllaceae are related. There is also consensus (Hutchinson dissenting) that the Polemoniaceae are related to the Solanaceae and Convolvulaceae. Opinion is divided as to whether the Boraginaceae have closer affinity with the Polemoniaceae or with the Labiatae. There is also consensus (Hutchinson dissenting) that the Labiatae and Verbenaceae are related. There is again consensus (Hutchinson again dissenting) that the Scrophulariaceae and Bignoniaceae are related.

GLOSSARY FOR POLEMONIALES

Polemoniales. Order in subclass Gamopetalidae.

Tubiflorae. Classis in cohors Gamopetalae (Endl.)

Polemoniales. Cohors in series Bicarpellatae (B.&H.) In subclassis Gamopetalae.

Tubiflorae. Unterreihe in Reihe Haplostemones (Eichl.) In Klasse Sympetalae.

Tubiflorae. Series in subclassis Metachlamydeae (Engl.)

Polemoniales. Order in superorder Sympetalae-Dicarpellatae (Bessey). In subclass Strobiloideae.

Polemoniales. Order in division Herbaceae (Hutch.)

Tubiflorae. Reihe in Unterklasse Sympetalae (Melch.)

Polemoniales. Order in subclass Asteridae (Cronq.)

Boraginales. Order in subclass Gamopetalidae.

Boraginales. Order in division Herbaceae (Hutch.)

Verbenales. Order in subclass Gamopetalidae.

Verbenales. Order in division Lignosae (Hutch.)

Labiatales. Order in subclass Gamopetalidae.

Nuculiferae. Classis in cohors Gamopetalae (Endl.)

Lamiales. Cohors in series Bicarpellatae (B.&H.) In subclass Gamopetalae.

Labiatiflorae. Unterreihe in Reihe Haplostemones (Eichl.) In Klasse Sympetalae.

Lamiales. Order in superorder Sympetalae-Polycarpellatae (Bessey). In subclass Strobiloideae.

Lamiales. Order in division Herbaceae (Hutch.)

Lamiales. Order in subclass Asteridae (Cronq.)

Solanales. Order in subclass Gamopetalidae.

Solanales. Order in division Herbaceae (Hutch.)

Scrophulariales. Order in subclass Gamopetalidae.

Personatae. Classis in cohors Gamopetalae (Endl.)

Personales. Cohors in series Bicarpellatae (B.&H.) In subclassis Gamopetalae.

Scrophulariales. Order in superorder Sympetalae-Dicarpellatae (Bessey). In subclass Strobiloideae.

Personales. Order in division Herbaceae (Hutch.)

Scrophulariales. Order in subclass Asteridae (Cronq.)

Bignoniales. Order in subclass Gamopetalidae.

Bignoniales. Order in division Lignosae (Hutch.)

Convolvulaceae. Family in order Polemoniales.

Convolvulaceae. Ordo in classis Tubiflorae (Endl.)

Convolvulaceae. Ordo in cohors Polemoniales (B.&H.)

Convolvulaceae. Familie in Unterreihe Tubiflorae (Eichl.)

Convolvulaceae. Familia in series Tubiflorae (Engl.)

Convolvulaceae. Family in order Polemoniales (Bessey), (Cronq.)

Convolvulaceae. Family in order Solanales (Hutch.)

Convolvulaceae. Familie in Reihe Tubiflorae (Melch.)

Polemoniaceae. Family in order Polemoniales.

Polemoniaceae. Ordo in classis Tubiflorae (Endl.)

Polemoniaceae. Ordo in cohors Polemoniales (B.&H.)

Polemoniaceae. Familie in Unterreihe Tubiflorae (Eichl.)

Polemoniaceae. Familia in series Tubiflorae (Engl.)

Polemoniaceae. Family in order Polemoniales (Bessey), (Hutch.), (Cronq.)

Polemoniaceae. Familie in Reihe Tubiflorae (Melch.)

Hydrophyllaceae. Family in order Polemoniales.

Hydrophylleae. Ordo in classis Tubiflorae (Endl.)

Hydrophyllaceae. Ordo in cohors Polemoniales (B.&H.)

Hydrophyllaceae. Familie in Unterreihe Tubiflorae (Eichl.)

Hydrophyllaceae. Familia in series Tubiflorae (Engl.)

Hydrophyllaceae. Family in order Polemoniales (Bessey), (Hutch.), (Cronq.)

Hydrophyllaceae. Familie in Reihe Tubiflorae (Melch.)

Boraginaceae. Family in order Polemoniales.

Asperifoliae. Ordo in classis Nuculiferae (Endl.)

Boragineae. Ordo in cohors Polemoniales (B.&H.)

Asperifoliae. Familie in Unterreihe Tubiflorae (Eichl.)

Borraginaceae. Familia in series Tubiflorae (Engl.)

Borraginaceae. Family in order Polemoniales (Bessey).

Boraginaceae. Family in order Boraginales (Hutch.)

Boraginaceae. Familie in Reihe Tubiflorae (Melch.)

Boraginaceae. Family in order Lamiales (Cronq.)

Verbenaceae. Family in order Polemoniales.

Verbenaceae. Ordo in classis Nuculiferae (Endl.)

Verbenaceae. Ordo in cohors Lamiales (B.&H.)

Verbenaceae. Familie in Unterreihe Labiatiflorae (Eichl.)

Verbenaceae. Familia in series Tubiflorae (Engl.)

Verbenaceae. Family in order Lamiales (Bessey), (Cronq.)

Verbenaceae. Family in order Verbenales (Hutch.)

Verbenaceae. Familie in Reihe Tubiflorae (Melch.)

Labiatae. Family in order Polemoniales.

Labiatae. Ordo in classis Nuculiferae (Endl.)

Labiatae. Ordo in cohors Lamiales (B.&H.)

Labiatae. Familie in Unterreihe Labiatiflorae (Eichl.)

Labiatae. Familia in series Tubiflorae (Engl.)

Lamiaceae. Family in order Lamiales (Bessey), (Hutch.)

Labiatae. Familie in Reihe Tubiflorae (Melch.)

Labiatae. Family in order Lamiales (Cronq.)

Solanaceae. Family in order Polemoniales.

Solanaceae. Ordo in classis Tubiflorae (Endl.)

Solanaceae. Ordo in cohors Polemoniales (B.&H.)

Solanaceae. Familie in Unterreihe Tubiflorae (Eichl.)

Solanaceae. Familia in series Tubiflorae (Engl.)

Solanaceae. Family in order Polemoniales (Bessey), (Cronq.)

Solanaceae. Family in order Solanales (Hutch.)

Solanaceae. Familie in Reihe Tubiflorae (Melch.)

Scrophulariaceae. Family in order Polemoniales.

Scrophularineae. Ordo in classis Personatae (Endl.)

Scrophularineae. Ordo in cohors Personales (B.&H.)

Scrophulariaceae. Familie in Unterreihe Labiatiflorae (Eichl.)

Scrophulariaceae. Familia in series Tubiflorae (Engl.)

Scrophulariaceae. Family in order Scrophulariales (Bessey), (Cronq.)

Scrophulariaceae. Family in order Personales (Hutch.)

Scrophulariaceae. Familie in Reihe Tubiflorae (Melch.)

Bignoniaceae. Family in order Polemoniales.

Bignoniaceae. Ordo in classis Personatae (Endl.)

Bignoniaceae. Ordo in cohors Personales (B.&H.)

Bignoniaceae. Familie in Unterreihe Labiatiflorae (Eichl.)

Bignoniaceae. Familia in series Tubiflorae (Engl.)

Bignoniaceae. Family in order Scrophulariales (Bessey), (Cronq.)

Bignoniaceae. Family in order Bignoniales (Hutch.)

Bignoniaceae. Familie in Reihe Tubiflorae (Melch.)

Oleaceae. See 38 4.

Oleaceae. Family in order Scrophulariales (Cronq.)

Plantaginaceae. See 41 4.

Plantagineae. Familie in Unterreihe Labiatiflorae (Eichl.)

41 1:1

PLANTAGINALES

The Plantaginales include only the Plantaginaceae, and in our area the family is represented almost exclusively by the genus <u>Plantago</u>. The flowers of the order are gamopetalous, tetracyclic, tetrameric, hypogynous, and in scapose spikes. The plants are herbaceous and acaulescent. Their leaves have parallel venation.

<u>Plantago</u> is the ancient Latin name for plantain. The name signifies: "plant which has leaves which resemble footprints." From its occurence in disturbed areas, the American Indians are said to have called plantain "white-man's foot."

41 2:1

PLANTAGINACEAE

dHypoG/2L1-4A(2-)4(epipet.alt.)C/4(scarious)K4/pyxis

lvs.basal

Dicotyledonous plants with hypogynous flowers. Gynoecium of 2 united carpels and 1 to 4 locules. Androecium of 4 stamens (or as few as 2) (epipetalous and alternate with the petals). Corolla of 4 united petals (scarious). Calyx of 4 sepals. Fruit a pyxis. Leaves basal.

Plantago sp.

dHypoG/2A4(epipet.alt.)C/4(scarious)K4/

PLANTAGINACEAE

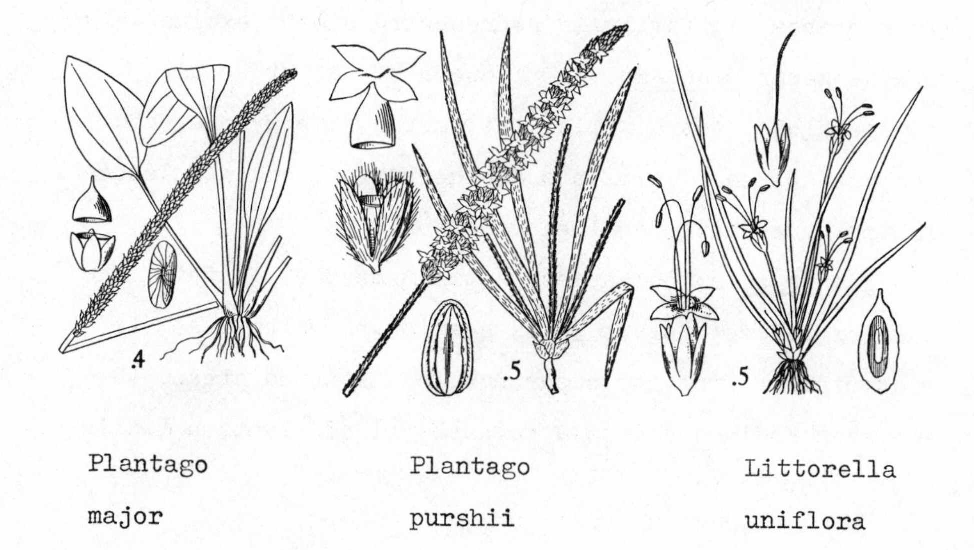

Plantago major

-

plantain

Plantago purshii

-

Littorella uniflora

-

41 3:1

Endlicher does not recognize the Plantaginales but places the Plantaginaceae in the Plumbaginales. Here the order precedes the Asterales. Bentham and Hooker place the Plantaginaceae in the Gamopetalidae, the second of their three dicot subclasses. Here the family is treated as anomalous and is assigned to no order. This is equivalent to the creation of an order Plantaginales for this single family. The family stands after the gamopetalous, tetracyclic, and hypogynous orders of the superorder Bicarpellatae: Gentianales, Polemoniales, Scrophulariales, Labiatales. Eichler does not recognize the Plantaginales. He places the Plantaginaceae within the Labiatales. Engler places the Plantaginales in the Gamopetalidae, the second of his two dicot subclasses. Here the order is associated in a gamopetalous, tetracyclic, and hypogynous order group: Gentianales, Polemoniales, Plantaginales. Bessey does not recognize the Plantaginales. He places the Plantaginaceae in the Primulales. An important character for this classification is the circumscissile capsule or pyxis fruit of the Plantaginaceae and of some members of the Primulaceae. Hutchinson places the Plantaginales in the Herbacidae, the second of his two dicot subclasses. Here the order stands immediately after the Primulales, and he diagrams as a phylogenetic sequence: Primulales, Plantaginales. Melchior retains the Plantaginales in their Englerian position. Cronquist places the Plantaginales in the gamopetalous and tetracyclic Asteridae, the sixth of his six dicot subclasses. Here the order is associated in a hypogynous order group: Gentianales, Polemoniales, Labiatales, Plantaginales, Scrophulariales. He diagrams the Plantaginales as without specific affinity to other members of this order group.

All of the systems accepting the Plantaginales restrict it to the single family Plantaginaceae.

41 4:1

GLOSSARY FOR PLANTAGINALES

Plantaginales. Order in subclass Gamopetalidae.

Ordo anomalis. In series Bicarpellatae (B.&H.) In subclassis Gamopetalae.

Plantaginales. Series in subclassis Metachlamydeae (Engl.)

Plantaginales. Order in division Herbaceae (Hutch.)

Plantaginales. Reihe in Unterklasse Sympetalae (Melch.)

Plantaginales. Order in subclass Asteridae (Cronq.)

Plantaginaceae. Family in order Plantaginales.

Plantagineae. Ordo in classis Plumbaginales (Endl.)

Plantagineae. Ordo anomalis (B.&H.)

Plantagineae. Familie in Unterreihe Labiatiflorae (Eichl.)

Plantaginaceae. Familia in series Plantaginales (Engl.)

Plantaginaceae. Family in order Primulales (Bessey).

Plantaginaceae. Family in order Plantaginales (Hutch.), (Cronq.)

Plantaginaceae. Familie in Reihe Plantaginales (Melch.)

RUBIALES

The Rubiales are gamopetalous, tetracyclic, and epigynous with epipetalous and mostly distinct stamens. The order contains the Rubiaceae and Caprifoliaceae. The Rubiaceae have usually a bicarpellary ovary, tetrameric floral structure, and stipulate leaves. The Caprifoliaceae have usually an ovary that has three to five carpels, floral structure which is pentameric, and estipulate leaves.

Rubia is the ancient Latin name of madder. The root of the name signifies "red," and the root of the plant yields a red dye. *Caprifolium* is a pre-Linnaean name of herbalists. It is composed of two elements meaning "goat" and "leaf," but the signification of the compound is obscure. The name is preserved in *Lonicera caprifolium*. Here nomenclatural purists capitalize the initial letter of the specific word in order to indicate that it was originally a generic name. The neuter ending *-um* does not agree with the feminine ending of *Lonicera*, and indicates that *caprifolium* is a nominative noun and not an adjectival or a genitive form.

RUBIACEAE

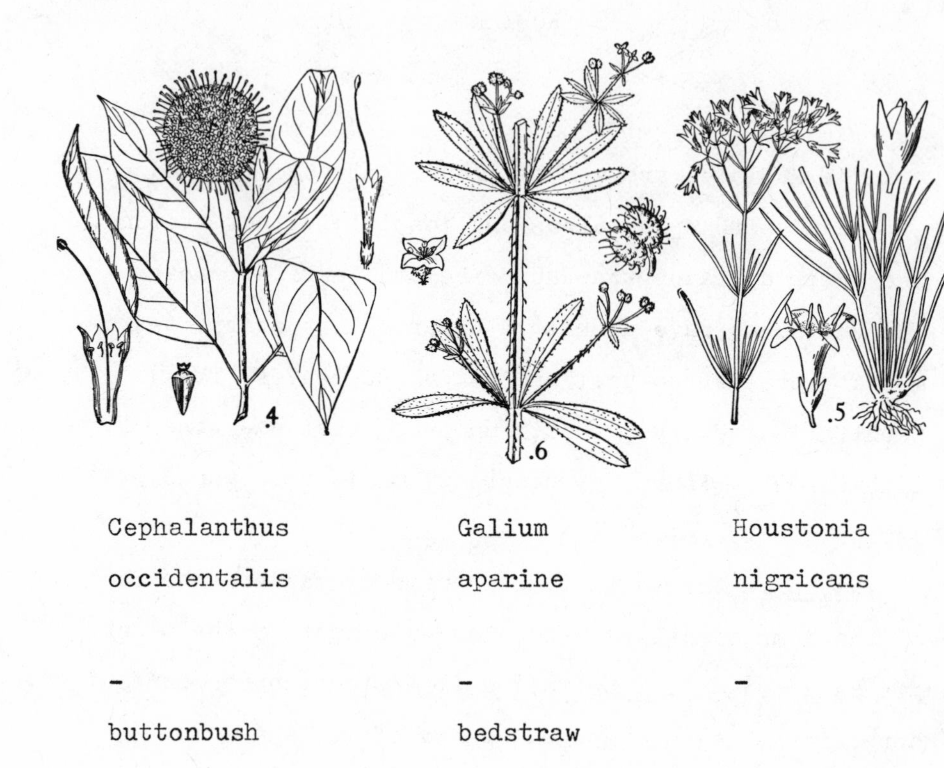

Cephalanthus occidentalis	Galium aparine	Houstonia nigricans
-	-	-
buttonbush	bedstraw	

RUBIACEAE

dEpiG̸2-4A(3-)4(epipet.alt.)C̸(3-)4Ki,±̸,4(-6),q/caps., ber.,drup.,lvs.opp.lvs.w.stipules

Dicotyledonous plants with epigynous flowers. Gynoecium of 2 to 4 united carpels. Androecium of 4 stamens (or as few as 3) (epipetalous and alternate with the petals). Corolla of 4 united petals (or as few as 3). Calyx of 4 sepals (or as many as 6), or the number not evident, the evident sepals free or more or less united. Fruit a capsule, berry, or drupaceous. Leaves opposite. Leaves with stipules.

Cephalanthus occidentalis

dEpiG/2A4(epipet.alt.)C/4K/4/

CAPRIFOLIACEAE

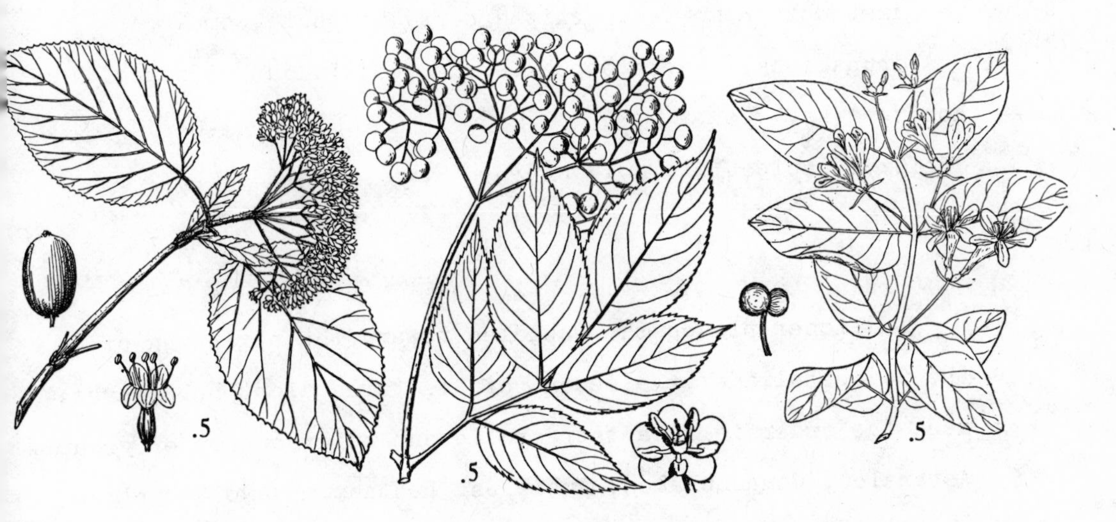

Viburnum lantana	Sambucus canadensis	Lonicera tatarica
wayfaring tree	-	Tartarian honeysuckle
-	elder	

CAPRIFOLIACEAE

dEpiGi̸2-5A4-5(epipet.alt.)Ci̸a,a̸,(3-)5Ki,i̸,(3-)5/ber.,caps., drup.,lvs.opp.lvs.not w.stipules

Dicotyledonous plants with epigynous flowers. Gynoecium of 2 to 5 united carpels. Androecium of 4 to 5 stamens (epipetalous and alternate with the petals). Corolla of 5 united petals (or as few as 3), actinomorphic or zygomorphic. Calyx of 5 sepals (or as few as 3), distinct or united. Fruit a berry, capsule, or drupaceous. Leaves opposite. Leaves without stipules.

Lonicera sp.

dEpiGi̸3A5(epipet.alt.)Ci̸a̸5Ki̸5/

42 3:1

Endlicher places the Rubiales toward the beginning of the Gamopetalidae, the second of his three dicot subclasses. Here the order is in a tetracyclic and epigynous order group: Asterales, Campanulales, Rubiales. Bentham and Hooker place the Rubiales at the beginning of the Gamopetalidae, the second of their two dicot subclasses. Here the order is in the tetracyclic and epigynous superorder Inferae: Rubiales, Asterales, Campanulales. Eichler places the Rubiales in the Gamopetalidae, the first of his two dicot subclasses. Here the order is in the tetracyclic superorder Haplostemones, and within this in an epigynous order group: Rubiales (incl. Compositae), Campanulales. Engler places the

Rubiales in the Gamopetalidae, the second of his two dicot subclasses. Here the order is in a tetracyclic and epigynous order group: Rubiales, Campanulales. Bessey places the Rubiales in the Dicotocotyloididae, the second of his two dicot subclasses. Here the order is in the gamopetalous, tetracyclic, and epigynous superorder Sympetalae: Rubiales, Campanulales, Asterales. Bessey diagrams as a phylogenetic sequence: Umbellales, Rubiales. From the Rubiales he separately derives the Campanulales and Asterales. Hutchinson places the Rubiales in the Lignosidae, the first of his two dicot subclasses. Here the order stands remote from the Araliales into which he puts the Caprifoliaceae. The Rubiales are in a gamopetalous and tetracyclic order group: Oleales, Apocynales, Rubiales, Bignoniales, Verbenales. Their epigynous character is exceptional in this group. Melchior does not recognize the Rubiales, and he includes the Rubiaceae in the Gentianales. His Caprifoliales, however, retain the position of the Englerian Rubiales. Cronquist includes the Rubiales and Caprifoliales as adjacent orders in the gamopetalous and tetracyclic Asteridae, the sixth of his six dicot subclasses. Here the order is in an epigynous order group: Campanulales, Rubiales, Caprifoliales, Asterales. He diagrams a phylogenetic line leading to the Rubiales. The Gentianales are an offshoot near its base, and the Caprifoliales and Asterales are offshoots from the Rubiales.

Endlicher, Eichler, Engler, and Bessey place the Rubiaceae ahead of the Caprifoliaceae. Bentham and Hooker reverse this order. Hutchinson, Melchior, and Cronquist place the Rubiaceae and Caprifoliaceae in separate orders. Cronquist, however, accepts a relationship between the Rubiaceae and Caprifoliaceae; whereas Hutchinson and Melchior hold these families to be unrelated.

42 4:1

GLOSSARY FOR RUBIALES

Rubiales. Order in subclass Gamopetalidae.

Caprifolia. Classis in cohors Gamopetalae (Endl.)

Rubiales. Cohors in series Inferae (B.&H.)

In subclassis Gamopetalae.

Aggregatae. Unterreihe in Reihe Haplostemones (Eichl.)

In Klasse Sympetalae.

Rubiales. Series in subclassis Metachlamydeae (Engl.)

Rubiales. Order in superorder Sympetalae (Bessey).

In subclass Cotyloideae.

Rubiales. Order in division Lignosae (Hutch.)

Rubiales. Order in subclass Asteridae (Cronq.)

Caprifoliales. Order in subclass Gamopetalidae.

Dipsacales. Reihe in Unterklasse Sympetalae (Melch.)

Rubiaceae. Family in order Rubiales.

Rubiaceae. Ordo in classis Caprifolia (Endl.)

Rubiaceae. Ordo in cohors Rubiales (B.&H.)

Rubiaceae. Familie in Unterreihe Aggregatae (Eichl.)

Rubiaceae. Familia in series Rubiales (Engl.)

Rubiaceae. Family in order Rubiales (Bessey), (Hutch.), (Cronq.)

Rubiaceae. Familie in Reihe Gentianales (Melch.)

Caprifoliaceae. Family in order Rubiales.

Lonicereae. Ordo in classis Caprifolia (Endl.)

Caprifoliaceae. Ordo in cohors Rubiales (B.&H.)

Caprifoliaceae. Familie in Unterreihe Aggregatae (Eichl.)

Caprifoliaceae. Familia in series Rubiales (Engl.)

Caprifoliaceae. Family in order Rubiales (Bessey).

Caprifoliaceae. Family in order Araliales (Hutch.)

Caprifoliaceae. Familie in Reihe Dipsacales (Melch.)

Caprifoliaceae. Family in order Dipsacales (Cronq.)

Compositae. See 44 4.

Compositae. Familie in Unterreihe Aggregatae (Eichl.)

43 1:1

CUCURBITALES

The order Cucurbitales includes the single family Cucurbitaceae. The Cucurbitaceae have a tricarpellate inferior ovary becomming in fruit a pepo, a fruit of berry type, having seeds attached to parietal placentae and a rind derived from tissues of the receptacle which surrounds the inferior ovary of the flower. They are vining plants with tendrils and broad, palmately-veined leaves.

Cucurbita is the ancient Latin word for "gourd," and the term gourd is itself derived through French from Latin cucurbita.

43 2:1

CUCURBITACEAE

dEpiG/3L1-3Ai,/,3(epipet.alt.)C/5(-6)K/5(-6)/monoec., dioec.,pepo vines lvs.w.venat.palmate

Dicotyledonous plants with epigynous flowers. Gynoecium of 3 united carpels and 1 to 3 locules. Androecium of 3 free, or united, stamens (epipetalous and alternate with the petals). Corolla of 5 united petals (or as many as 6). Calyx of 5 united sepals (or as many as 6). Plants monoecious, or dioecious.

CUCURBITACEAE

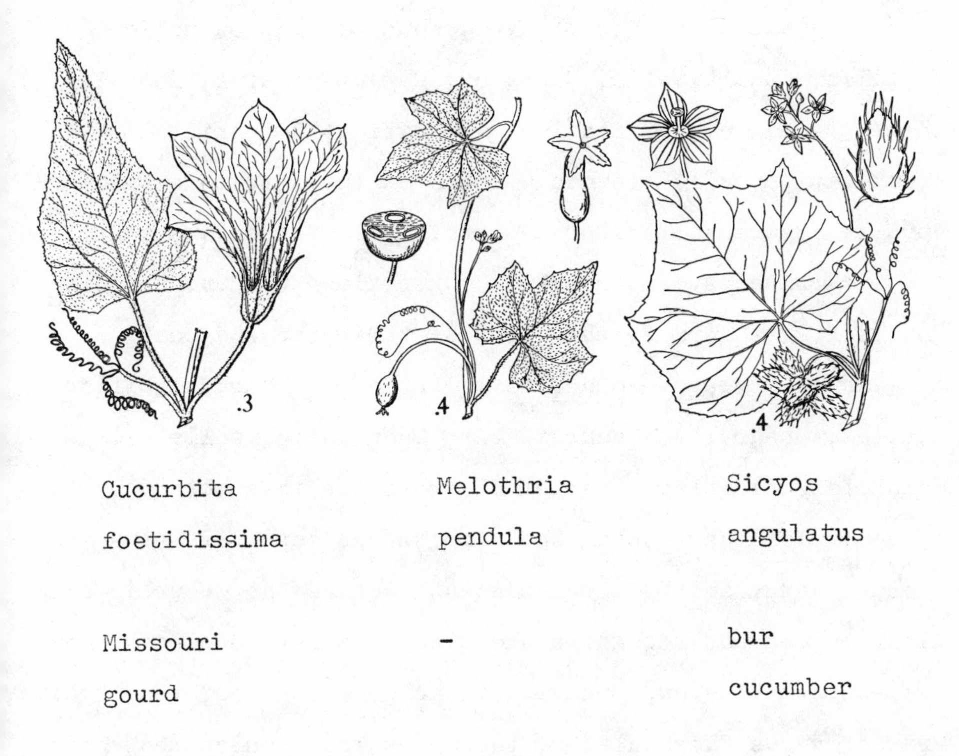

Cucurbita foetidissima	Melothria pendula	Sicyos angulatus
Missouri gourd	-	bur cucumber

Fruit a pepo. Plants vines. Leaves with venation palmate.

Cucurbita foetidissima

dEpiGi̸3L1A¯3(epipet.alt.)Ci̸5Ki̸5/monoec.

Endlicher places the Cucurbitales in the Polypetalidae, the third of his three dicot subclasses. Here the order is in a thealian order group, hypogynous but not eucyclic: Violales, Cucurbitales, Cactales, Caryophyllales, Malvales, Theales. This classification gives weight to parietal placentation while avoiding stress on the sympetalous corolla, tetracyclic floral parts, and inferior ovary of the Cucurbitales. Bentham and Hooker place the Cucurbitales in the Polypetalidae, the first of their three dicot subclasses. Here the order is in the superorder Calyciflorae: Rosales, Onagrales, Cucurbitales, Cactales, Umbellales. This classification gives emphasis to the inferior ovary while avoiding stress on the sympetalous corolla, tetracyclic floral parts, and parietal placentation of the Cucurbitales. Eichler does not recognize the Cucurbitales. He places the Cucurbitaceae as an appendix to the gamopetalous, tetracyclic, and epigynous Campanulales, the appendicular placement being indicative of diffidence occasioned by the parietal placentation and vining habit of the family. Engler and Prantl do not recognize the Cucurbitales, but following Eichler they place the Cucurbitaceae in the Campanulales. Engler and Gilg, however, raise the Cucurbitaceae to ordinal status; and here the Rubiales, Cucurbitales, and Campanulales are in a gamopetalous, tetracyclic, and epigynous order group. Bessey places the Cucurbitales in the Dicotocotyloididae, the second of his two dicot subclasses.

Here the order is in the superorder Apopetalae: Rosales, Onagrales, Cucurbitales, Cactales, Celastrales, Sapindales, Umbellales. He diagrams as a phylogenetic sequence: Rosales, Onagrales, Cucurbitales. Hutchinson places the Cucurbitales in the Lignosidae, the first of his two dicot subclasses. He diagrams the Cucurbitales at the terminus of a phylogenetic line originating with antecedents of the Theales and having an offshoot leading to the Cactales. Melchior places the Cucurbitales in the Agamopetalidae, the first of his two dicot subclasses. Here the order is in the order group: Malvales, Elaeagnales, Violales, Cucurbitales. Cronquist does not recognize the Cucurbitales, but he places the Cucurbitaceae in his Violales.

There are three categories of interpretation for the Cucurbitales, which may be characterized as (1) thalamifloric, (2) calycifloric, and (3) gamopetalidic. The first interpretation derives the group from or associates it with polypetalous and hypogynous taxa of thealian or violalian affinity. Supporters of this view are Endlicher, Hutchinson, Melchior, and Cronquist. The second interpretation derives the group from or associates it with polypetalous perigynous and polypetalous epigynous taxa of rosalian or onagralian affinity. Supporters of this view are Bentham and Hooker, and Bessey. The third interpretation derives the group from or associates it with gamopetalous, tetracyclic, and epigynous taxa of campanulalian affinity. Supporters of this view are Eichler and Engler. The first interpretation, which is also the oldest of those here represented, predominates at present.

43 4:1

GLOSSARY FOR CUCURBITALES

Cucurbitales. Order in subclass Gamopetalidae.
  Peponiferae. Classis in cohors Dialypetalae (Endl.)
  Passiflorales. Cohors in series Calyciflorae (B.&H.)
    In subclassis Polypetalae.
  Loasales. Order in superorder Apopetalae (Bessey).
    In subclass Cotyloideae.
  Cucurbitales. Order in division Lignosae (Hutch.)
  Cucurbitales. Reihe in Unterklasse Archichlamydeae
    (Melch.)

Cucurbitaceae. Family in order Cucurbitales.
  Cucurbitaceae. Ordo in classis Peponiferae (Endl.)
  Cucurbitaceae. Ordo in cohors Passiflorales (B.&H.)
  Cucurbitaceae. Familie in Unterreihe Campanulinae (Eichl.)
  Cucurbitaceae. Familia in series Campanulatae (Engl.)
  Cucurbitaceae. Family in order Loasales (Bessey).
  Cucurbitaceae. Family in order Cucurbitales (Hutch.)
  Cucurbitaceae. Familie in Reihe Cucurbitales (Melch.)
  Cucurbitaceae. Family in order Violales (Cronq.)

CAMPANULALES

The Campanulales are gamopetalous, tetracyclic, and epigynous with stamens coherent or connate. The order includes the Campanulaceae and Compositae. The Campanulaceae have flowers solitary or in various types of inflorescence, but not in involucrate heads, a distinctive characteristic of the Compositae. In most of the systems here considered *Lobelia* is included in the Campanulaceae, but the genus is placed in a separate family by Endlicher, Eichler, and Hutchinson. Bessey does not recognize the Compositae but raises the taxa usually treated as tribes of the Compositae to familial status.

Campanula is derived from a medieval Latin name meaning "little bell," and its botanical signification is: "plant having flowers with corolla resembling a small bell." *Compositae* is the feminine plural form of the participle *compositus* of Latin *componere* "to place together." Etymologically it signifies: "plants with floral elements aggregated into a compound structure," which resembles an individual flower in function and in superficial appearance.

CAMPANULACEAE

Lobelia cardinalis	Campanula rapunculoides	Specularia perfoliata
-	-	Venus' looking-glass

CAMPANULACEAE

dEpiG/2-5Ai,_,‾,5(alt.,epipet.alt.)C/a,/a,5K_(3-)5/caps.

Dicotyledonous plants with epigynous flowers. Gynoecium of 2 to 5 united petals. Androecium of 5 distinct stamens, or the stamens united at the base or by the anthers (the stamens alternate with the petals, or epipetalous and alternate with the petals). Corolla of 5 united petals, actinomorphic or zygomorphic. Calyx of 5 sepals (or as few as 3) united at the base. Fruit a capsule.

Campanula sp.

dEpiGi̸3A5Ci̸5K_5/

Lobelia sp.

dEpiGi̸2A¯5(epipet.alt.)Ci̸d̸5(2+3)K_5/

Corolla of 5 petals, united and zygomorphic (the lobes in groups of 2 and of 3 members).

COMPOSITAE

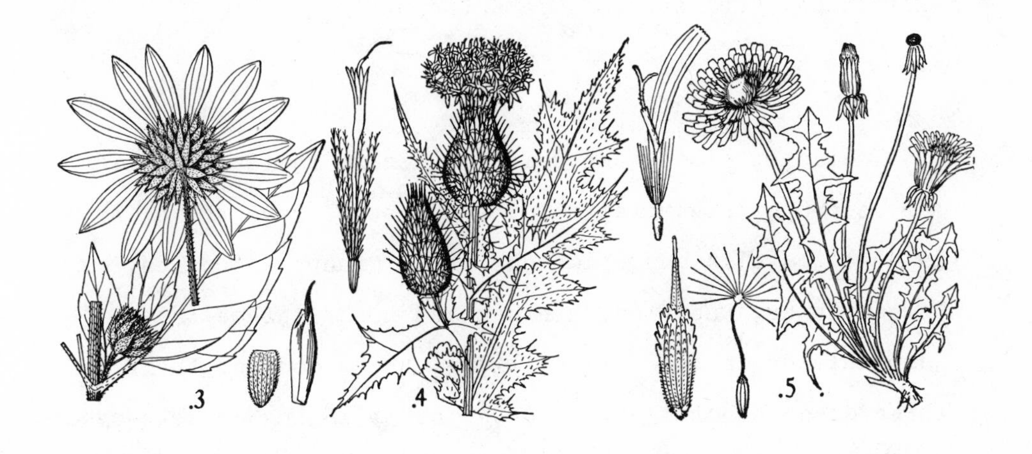

Helianthus annuus	Cirsium vulgare	Taraxacum officinale
common sunflower	bull thistle	- dandelion

COMPOSITAE

dEpiG/2L1A‾5(epipet.alt.)C/a,a/,(2-)5K2-n(pappus),0/perf., unisex.,ach.

Dicotyledonous plants with epigynous flowers. Gynoecium of 2 united petals and a single locule. Androecium of 5 stamens united by the anthers (epipetalous and alternate with the petals). Corolla of 5 united petals (or as few as 2), actinomorphic or zygomorphic. Calyx of 2 to many parts (a pappus), or none. Flowers perfect, or unisexual. Fruit an achene.

Helianthus sp. (tubular flower)

dEpiG/2L1A‾5(epipet.alt.)C/a5K2(pappus)/

44 3:1

Endlicher places the Campanulales in the Gamopetalidae, the second of his three dicot subclasses. Here the order is in a tetracyclic and epigynous order group: Asterales, Campanulales, Rubiales. Bentham and Hooker place the Campanulales in the Gamopetalidae, the second of their three dicot subclasses. Here the order is in the tetracyclic and epigynous superorder Inferae: Rubiales, Asterales, Campanulales. Eichler places the Campanulales in the Gamopetalidae, the first of his two dicot subclasses. Here the order is in the tetracyclic superorder Haplostemones, and within this in the epigynous order group: Rubiales (incl. Compositae), Campanulales. Engler classifies the

Campanulales in the Gamopetalidae, the second of his two dicot subclasses. Here the order stands at the end of the tetracyclic and epigynous order group: Rubiales, Campanulales. Bessey classifies the Campanulales in the Dicotocotyloididae, the second of his two dicot subclasses. Here the order is in the gamopetalous, tetracyclic, and epigynous superorder Sympetalae: Rubiales, Campanulales, Asterales. Bessey diagrams the Campanulales and Asterales as separately derived from the Rubiales. Hutchinson places the Campanulales in the Herbacidae, the second of his two dicot subclasses. Here the order is in the gamopetalous, tetracyclic, and epigynous order group: Campanulales, Asterales. He diagrams as a phylogenetic sequence: Saxifragales, Campanulales, Asterales. Melchior retains the Campanulales in their Englerian position. Cronquist places the Campanulales in the Asteridae, the sixth of his six dicot subclasses. Here it stands in a gamopetalous, tetracyclic, and epigynous order group: Campanulales, Rubiales, Caprifoliales, Asterales. He diagrams a bifurcating phylogenetic line. The first branch leads to polemonialian orders, and an offshoot from it leads to the Campanulales. The second branch leads to the Rubiales, and from this order arise separate lines leading to the Asterales and Caprifoliales.

There is agreement that the Campanulales should be recognized as an order. Eichler alone includes the

Cucurbitaceae in the order. Engler and Melchior include the Compositae in the order. Eichler alone classifies the Compositae in the Rubiales. Endlicher, Bentham and Hooker, Hutchinson, and Cronquist recognize the Asterales as containing the Compositae. Bessey also recognizes the Asterales, but he places at familial level the taxa which in other systems are tribes of the Compositae. There is then a preponderance of opinion recognizing the Asterales in addition to the Campanulales.

44 4:1

GLOSSARY FOR CAMPNULALES

Campanulales. Order in subclass Gamopetalidae.

Campanulinae. Classis in cohors Gamopetalae (Endl.)

Campanales. Cohors in series Inferae (B.&H.)
In subclassis Gamopetalae.

Campanulinae. Unterreihe in Reihe Haplostemones (Eichl.)
In Klasse Sympetalae.

Campanulatae. Series in subclassis Metachlamydeae (Engl.)

Campanulales. Order in superorder Sympetalae (Bessey).
In subclass Cotyloideae.

Campanales. Order in division Herbaceae (Hutch.)

Campanulales. Reihe in Unterklasse Sympetalae (Melch.)

Campanulales. Order in subclass Asteridae (Cronq.)

Asterales. Order in subclass Gamopetalidae.

Aggregatae. Classis in cohors Gamopetalae (Endl.)

Asterales. Cohors in series Inferae (B.&H.)

In subclassis Gamopetalae.

Asterales. Order in superorder Sympetalae (Bessey).

In subclass Cotyloideae.

Asterales. Order in division Herbaceae (Hutch.)

Asterales. Order in subclass Asteridae (Cronq.)

Campanulaceae. Family in order Campanulales.

Campanulaceae. Ordo in classis Campanulinae (Endl.)

Campanulaceae. Ordo in cohors Campanales (B.&H.)

Campanulaceae. Familie in Unterreihe Campanulinae (Eichl.)

Campanulaceae. Familia in series Campanulatae (Engl.)

Campanulaceae. Family in order Campanulales (Bessey), (Cronq.)

Campanulaceae. Family in order Campanales (Hutch.)

Campanulaceae. Familie in Reihe Campanulales (Melch.)

Compositae. Family in order Campanulales.

Compositae. Ordo in classis Aggregatae (Endl.)

Compositae. Ordo in cohors Asterales (B.&H.)

Compositae. Familie in Unterklasse Aggregatae (Eichl.)

Compositae. Familia in series Campanulatae (Engl.)

Asteraceae. Family in order Asterales (Hutch.), (Cronq.)

Compositae. Familie in Reihe Campanulales (Melch.)

Cucurbitaceae. See 43 4.

Cucurbitaceae. Familie in Unterreihe Campanulinae (Eichl.)

INDEX TO GLOSSARIES

The present survey includes fifty glossaries to the higher nomenclature of the eight systems of plant classification considered in the present survey. This index covers these glossaries. When more than one page number is given in an index entry, a principal page number is often distinguished from the others.

GENERAL INDEX AND VOCABULARY

This index includes entries for families, orders, and higher taxa. It also includes entries for the authors of the systems here considered and for their writings which contain the systems. It further includes a variety of other entries, including the terms on which information is provided below.

Most of the terminology here employed is explained not only in the many botanical glossaries but also in the standard dictionaries. A few terms, however, require special notice. In certain systems, a series of orders may be referred to as an _order group_ to indicate that the series is analogous to formal higher taxa of some other systems. In connection with plant formulae the term _perianth_ is applied in cases where differentiation between corolla and calyx is lacking or where the floral envelope consists of a single whorl. A member of a perianth is in these cases termed a _tepal_. _Pseudanthium_ is a term for a grouping of flowers which together resemble a single flower. It occurs in current usage, but it is not included in older glossaries and dictionaries. _Psilogynous_ is a term here used coordinately with _hypogynous_, _perigynous_, and _epigynous_ for flowers consisting of gynoecium only, and therefore falling outside the scope of the latter three terms.

The area designated by the expression "Central and Northeastern United States" includes Minnesota, Iowa, Missouri, Kentucky, Virginia, and all the states northeast of the boundary which these states form.